REFLECTIONS *of* OUR PAST

REFLECTIONS *of* OUR PAST

How Human History Is Revealed in Our Genes

John H. Relethford

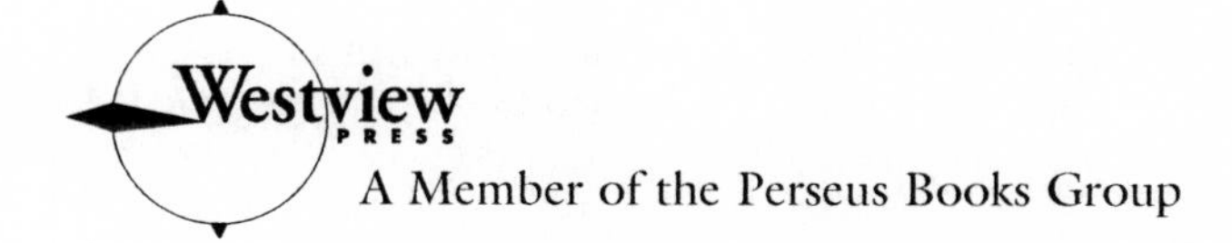

Westview Press books are available at special discounts for bulk purchases in the United States by corporations, institutions, and other organizations. For more information, please contact the Special Markets Department at the Perseus Books Group, 11 Cambridge Center, Cambridge MA 02142, or call (617) 252-5298 or (800) 255-1514 or email j.mccrary@perseusbooks.com.

Hardcover edition first published in 2003 in the United States of America by Westview Press, 5500 Central Avenue, Boulder, Colorado 80301–2877, and in the United Kingdom by Westview Press, 12 Hid's Copse Road, Cumnor Hill, Oxford OX2 9JJ

Paperback edition first published in 2004 by Westview Press.

Find us on the World Wide Web at www.westviewpress.com

Set in 10.5 point Galliard by the Perseus Books Group

A Cataloging-in-Publication data record for this book is available from the Library of Congress.
ISBN 0-8133-4259-7 (paperback)

The paper used in this publication meets the requirements of the American National Standard for Permanence of Paper for Printed Library Materials Z39.48-1984.

To my teachers and students,
for their inspiration,
And to my wife, Hollie, and
my sons, David, Ben, and Zane,
for their love

CONTENTS

A TIME MACHINE

I've often wished I had a time machine.

Not a machine that allowed me to travel through time (although I've entertained that fantasy as well), but some device that would allow me to peer into the past to see what happened. This desire emerged from reading time travel stories in comic books and science fiction novels as a youth and was deepened by my professional interest in the history of the human species. Where did we come from? Where and when did our ancestors live? When did humans first inhabit different parts of the world? These and other questions form the framework of much of my research. Because of this interest in the past, these questions are essentially historical in nature.

When you think of historical research, what types of information come to mind? There are a number of ways to obtain information from the past. For very recent times, we have the technology to actually see or hear past events that have been recorded on video and sound recorders. For earlier times, we have images from photographs, paintings and drawings, and sculpture. Written records, ranging from official documents to memoirs to diaries provide another window on history. Oral histories, including genealogies, stories, and legends, provide even more information.

There is an obvious limit to these methods. Video and audio technologies are recent inventions and cannot be used to see or hear events earlier in history. Written records are obviously limited to the time since humans developed writing systems (within the past 10,000 years at most). Oral histories are often subject to rapid change over time, as facts become myths and vice versa. To probe deeper and more fully into the past, we rely on inferences drawn from archaeological evidence, hoping to extract information about where, when, and how people lived from the remains of their past societies.

There is another source of information about the past: the genes that we all carry inside of us. Although we may lack direct historical or prehistorical documentation of past events, a record is present in our genes.

Each of us carries pieces of history in our genes. This book is about the search for human history not in documents or records, or even in ancient archaeological remains, but in the genetic material that we all carry. What can genetics tell us about our past?

This book examines how scientists use genetic information to reveal reflections of our past. My purpose is to provide the reader with information on not only *what* geneticists and anthropologists can tell about our past from genetic data but also *how* they do this. Genetics provides us with some interesting answers to our ancestry, and this book summarizes some of this exciting research.

I've structured this book in terms of a chronological journey, starting millions of years in the past with the origin of the first humanlike creatures and working forward to the present in each chapter. Not every time period or part of the world is covered; instead, the purpose is to give you an idea of some key events at different parts of our species' past. My goal here is to provide a broad picture of human history and prehistory illustrated by specific examples from different times and places. Some of the stops along the way include the following:

- Who are our closest living relatives, and when did we diverge along a different evolutionary path? (6–7 million years ago)
- When and where did modern humans first appear, and how are they related to other early humans? (130,000–2 million years ago)
- What happened to the Neandertals? (28,000–150,000 years ago)
- Where and when did the first Americans come from? (15,000–20,000 years ago)
- How did farming spread across Europe in prehistoric times? (6,000–10,000 years ago)
- When did humans first voyage into the Pacific, and where did they come from? (3,500 to 6,000 years ago)
- What is the genetic history of the Jews? (4,000 years ago)
- What was the genetic impact of the Viking invasions of Ireland? (1,200 years ago)
- What are the genetic roots of African Americans? (400 years ago)

Although my childhood dream of a time machine remains unfulfilled, the developments in genetics over the past century have provided another way to glimpse some of the history of our species.

I thank all of my colleagues who over the years have worked with me and helped me in my career. I am particularly grateful to those who have had the greatest impact on my research and professional development—John Blangero, Mike Crawford, Henry Harpending, Lyle Konigsberg, Frank Lees, Dick Wilkinson, and Milford Wolpoff.

I am very grateful to my sponsoring editor, Karl Yambert, for his assistance throughout all stages of this project. Karl responded with enthusiasm to my early, rather poorly conceived ideas for this book and helped me transform these ideas into reality. He also suggested the analogy of the palimpsest used in Chapter 5. I also thank Barbara Greer, project editor, for her professionalism and attention to details, and Jennifer Swearingen, copy editor, for a wonderful job in straightening out my tangled prose. I am also grateful to Tad Schurr for his review and recommendations.

JOHN H. RELETHFORD
December 2002

ONE

The History in Our Genes

I probably shouldn't admit this, but I spent a fair amount of time watching television while in graduate school.

Actually, "watching" may be an overstatement. Most of the time I simply had the television on as background noise, with little actual attention to the shows. I work best with some small amount of noise in the background. Thus, I would spend time in my apartment reading or writing, with occasional brief attention to the latest episode of *Happy Days* or *Dallas*.

Occasionally, a show would be aired that drew my primary attention away from my work and to the tube. In some cases, the show would be so interesting that my work would be put aside and I would give full attention to the show. One show that particularly captured my interest and attention, to the detriment of my studies for a week, was the miniseries *Roots,* which aired in late January 1977 on the ABC network. Based on the book by the same name, authored by Alex Haley in 1976, the miniseries portrayed the history of an African American family's ancestry. Haley's research started with bits of oral history passed down, generation to generation, which he supplemented with archival research. His search culminated with a trip to Africa, where he found corroborating oral histories establishing his ancestor as one Kunta Kinte, who as a young man had been captured by slavers and brought to the United States.

I was captivated by both the miniseries and the book. In addition to the sheer drama, I found something very appealing about the idea of finding one's "roots" and extending family history into the past. A number of years later, I acted upon this interest and contacted relatives on both my mother's side and my father's side of the family and found that in both cases someone had already compiled a fairly complete family tree. At periodic intervals over the years, I have added to this compilation and have been able to trace my family history as far back as an ancestor born in England in 1595.

In Search of History

Many of you may have similar interests in family genealogy. It is natural to be curious about where you come from. I also find it fascinating to consider that every person we know or meet also has a past and that we all have common connections at points in the past, either recent or ancient. When I pass someone in the store or in an airport, I often wonder how we might be related. Is that stranger a long-distant cousin? Did our ancestors know one another? Do we have similar roots, either in recent history or perhaps hundreds of generations in the past?

My specific interest in my family history grew into a broader interest in history and ancestry in general. Although I was developing these personal interests in graduate school, it did not occur to me then that much of my eventual professional research would ultimately revolve around "roots" and human history.

My specialization in graduate school was in the field of biological anthropology. Anthropology, as practiced in the United States, is a broadly based discipline focusing on the scientific study of humanity. Since we humans are by nature both biological and cultural organisms, anthropology looks at humanity from both biological and cultural perspectives. There are four different specializations within anthropology: cultural anthropology (the comparative study of behavior in living humans), linguistic anthropology (the comparative study of human languages), archaeology (the comparative study of behavior in historic and prehistoric societies), and biological anthropology. Biological anthropologists study the evolution and biological diversity of humans (and our close nonhuman relatives) both in the past and in the present. Some biological anthropologists specialize in the fossil record of human evolution, whereas others focus on the anatomy, evolution, and behavior of our close relatives, the nonhuman primates. Still others examine changes in human biology in response to changing environmental conditions, such as the impact of cultural changes on disease rates. My own interests focused on a field often known as *anthropological genetics*, which attempts to understand the factors that influence genetic variation in human populations. One of the areas studied by anthropological genetics is the relationship between population history and genetic variation, the subject of this book.

My primary research during graduate school focused on the genetics of Irish populations. I analyzed data on the anthropometrics (measurements of the body, face, and head) of adults from twelve towns in the western part of Ireland, trying to detect which populations were more similar physically to others and why. I was interested in the geographic, cultural, and demographic factors that could affect the pattern of similarity among these towns. For example, did the geographic distance between groups affect how similar they were? (Yes—towns closer together in space tend to be more similar physically, presumably because of shared genes due to migration).

Although my studies involved some background research on the history of Ireland, and of western Ireland in particular, my dissertation research did not incorporate population history to a large extent. This evolved later in my career. As a result of background reading for my dissertation, I had become aware of a collection of anthropometric data originally published in the 1890s. At the time, I was interested in these data simply for comparison with the populations in my dissertation research. However, upon completing my analyses, I found that, contrary to my dissertation thesis, the influence of geography on the similarity of populations was minimal and that the history of the region provided a better explanation of the biological similarity of these populations. Specifically, I found that two island populations off the west coast were rather distinct physically from populations on the west coast of Ireland and were in fact more similar to English populations. The history of these islands provided a clue; both had experienced an influx of English soldiers several centuries earlier to protect the Irish coast from pirates and possible invaders from continental Europe.

I describe this study in more detail in Chapter 9, but for the moment, I use it as a brief example of how the history of human populations is often reflected in their genetic makeup, be it in anthropometrics, blood groups, or DNA sequences. Following this study, I became interested in a broader picture of the relationship between history and biological variation in Ireland. I eventually conducted an analysis of anthropometric data from all of Ireland, which showed the biological impact of past invasions and settlements, primarily from Viking invasion and population settlement from England. In other words, these past events, most often analyzed in terms of their social and political effects, also left a pronounced and discernible biological impact.

Since the early 1990s, much of my research has broadened even further to consider the relationship between history and genetics for our entire species. One of the most interesting questions in anthropology today is the continuing debate over our species' "recent" evolutionary history—that is, within the past few hundred thousand years. This debate concerns where our ancestors lived 150,000 years ago. According to one model, *all* of our ancestors lived in Africa at that time. This model proposes that modern humans arose as a new species in Africa and then spread out across the world over the past 100,000 years or so. What makes this model fascinating to contemplate is that we know from the fossil record that there were already earlier humans living outside of Africa. If this model is correct, then these earlier humans were replaced by the newer humans dispersing from Africa. The assumption in this model is that all living humans share genes that originated in the recent past in Africa.

Not everyone agrees with this scenario. Others suggest that 150,000 years ago *some* of our ancestors lived in Africa, but many others lived outside of Africa, and there was never any dramatic replacement. If this hypothesis is true, then a critical question concerns not the replacement of one kind of human by another but the relative genetic contributions made by humans in different geographic regions in the past to the gene pool of living humans. Although the debate over modern human origins focuses largely on the fossil evidence, many studies have looked at the genetics of our present-day species for clues about what happened in the past. The basic premise is that whatever happened long ago left a discernible genetic impact in living humans. The debate on the origin of modern humans is described in more detail in Chapters 3 and 4. I raise it at this point to provide another example of the focus of this book—the use of genetics to reconstruct the history of human populations.

It should be apparent that I am using "history" in the broad sense of the word, meaning "whatever happened in the past." This view of history deals with more than just written history or recent times. The focus here is on the evolutionary history of a population, or a group of populations, or even an entire species. The scope of this book ranges from events that happened millions of years ago to events that occurred within the past few centuries. In a broad sense, this book deals with questions common to all humans—who are we, and where do we come from? In other words, what are our "roots"? The problem with much genealogical research is the limited time depth. The farther we go back in time, the less information we have.

Think about your own roots. How much information do you have about your ancestors? As you travel mentally back in time, you will likely see the quantity of information decrease sharply after a few generations. The first obvious step is your parents. In many cases, you will have a fair amount of information on them, perhaps where and when they were born, where they went to school, and the names of members in their immediate family. Most likely, you also have a lot of other information based on your life with them, including knowledge of their hobbies, favorite foods, and their dreams and fears.

How about your grandparents? Again, this varies, depending on whether you knew them personally and on the amount of contact you had with them. You might know one or both sets of grandparents very well, or not that well, but most (not all) of us at least know their names. Push it back a generation later. Do you know the names of all of your eight great-grandparents? I don't. Even with my interest in family genealogy, I know the names of only six out of eight. How about farther back—do you know the names of your sixteen great-great-grandparents? Do you actually know *how many* ancestors you have four generations ago? Sixteen great-great-grandparents is actually the *maximum* number of possible ancestors four generations in the past. If there was a recent ancestor in common, then you might have only fifteen distinct individuals among your sixteen great-grandparents (see Figure 1.1 for an example). Although I am fortunate to have inherited family genealogical data, I don't know the names of most of my ancestors beyond a few generations. I am luckier than most since my mother's family did not move around much over the past few centuries. Even given this information, the percentage of my known ancestors decreases significantly with each generation (Figure 1.2). After seven generations, I know less than 5 percent of my ancestors' names, let alone anything else about them. In some cases, I know dates of birth and death, but in many cases, I have only a name, and even some of these are marked as questionable in the family records. By eleven generations in the past, I have very little information; I know the names of only 2 out a maximum of 2,048 ancestors, which is less than 0.1 percent.

Unless you are very lucky, your list of known ancestors probably also shows a pattern of decreasing information after only a few generations. Only in rare cases, such as royalty, are we likely to know much more, and

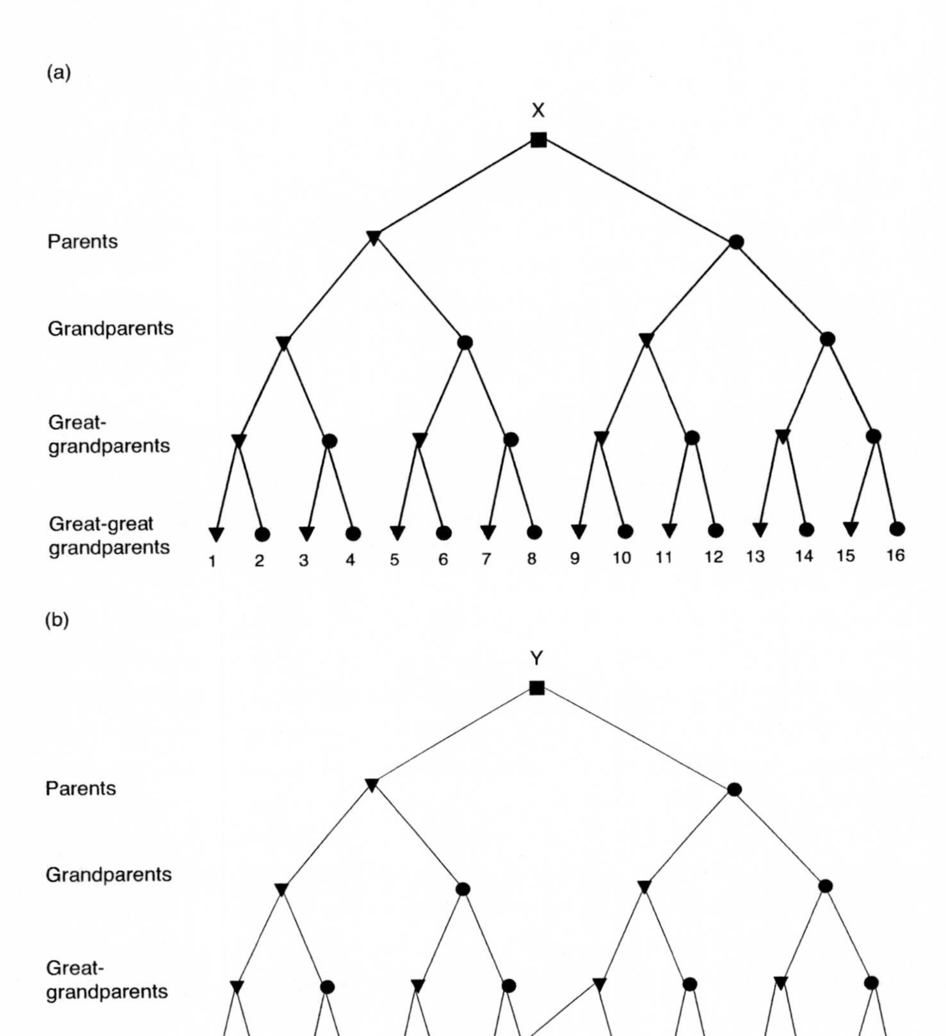

Figure 1.1 How many great-great-grandparents do persons X and Y have? (a) Person X has 2 parents, 4 grandparents, 8 great-grandparents, and 16 great-great-grandparents. This is the maximum number of ancestors four generations in the past. Triangles=males; circles=females (b) In this example, person Y has only 15 great-great-grandparents because the ancestor labeled "7" had children with two different women (persons "8" and "9"). In this case, the number of ancestors four generations in the past is less than the maximum because of distant inbreeding; Y's parents are half-second cousins who share 1 of 2 great-grandparents.

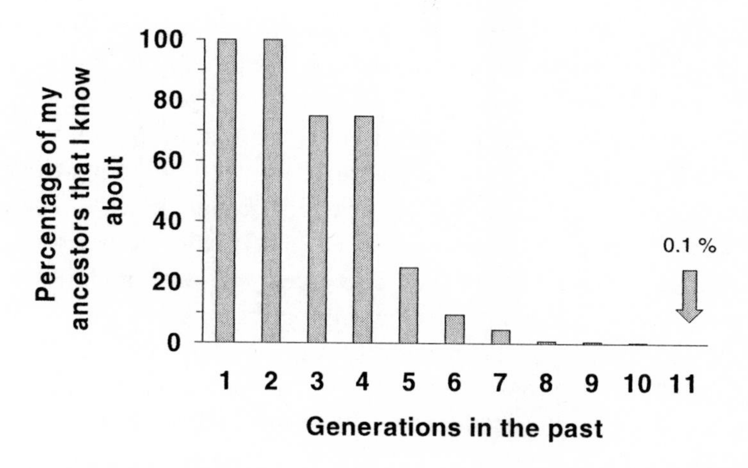

Figure 1.2 What I know about my ancestry. This graph plots the percentage of ancestors that are known for each of the past eleven generations. Even with family genealogies to guide me, the percentage of ancestors that can be identified decreases sharply over time.

even here, there is a definite limit to our written records. No one knows the name of even one ancestor living 10,000 years ago. Although the number of maximum ancestors increases the farther we go back in time, the chances that information about those ancestors exists, in either written or oral form, decline sharply. If we consider the limits on our direct genealogical information in terms of the long span of human history (in the broad sense), it becomes apparent that we know very, very little about our ancestors using a traditional genealogical approach. Even in cases where family records might go back for several centuries, this time depth is but a small fraction of the history of human beings, which extends back 2 million years (as judged by the first appearance of large-brained tool-users; see Chapter 3).

Consider a related question: *Where* do your ancestors come from? Again, we seldom have specific information for more than a few generations. In my case, all of my known ancestors on my mother's side for the past nine generations are listed as having been born in the United States, and I have no record of any immigration later than the mid-1600s. I have some records on my father's side going back in the United States to the mid-1700s, but no record of where any of them came from. Keep in mind that this information refers only to my *known* ancestors, which, as shown earlier, is a fraction of the total number of ancestors as I trace farther back in time. Questions of ancestry, both for individuals and for populations, can be answered using conventional historical data, but obviously for only limited periods of time.

Thinking about ancestry can be extremely frustrating. Most of us know our most immediate ancestors (our parents), most of whom knew their parents, and so on into the past. Over long periods of time, however, this information gets lost, for one reason or the other. Genetics provides a way of uncovering some of this information. Past events—from migration of people from one group to another to changes in population size—may have left a record behind in our genes. Our written and oral histories are incomplete and lack much time depth, but we carry a genetic signature of past events. In this sense, the study of genetics in living people can provide clues to past human history. As we study patterns of genetic variation, we look for such clues that have been preserved in our genes, generation to generation. We can (within limits) learn about the past by studying the present. As noted in the title of this book, genetics provides *reflections of our past.* This book deals with the search for human history using genetic data. By looking at the current patterns of genetic diversity of humans, we can reconstruct the past.

Genetics and Human History

As noted above, I take a very broad view of human history in this book, ranging from events that took place millions of years ago to events that have occurred in the past few centuries. I recall with some amusement reading world history texts that began with a section on "ancient" history, which often focused on Greece 2,500 years ago. To an anthropologist, this is very recent history when considering what we know archaeologically about human antiquity. Ancestors that can broadly be considered "human" have been around for two million years. Anatomically modern humans have been around for almost 150,000 years. The Agricultural Revolution that changed the nature of human existence began about 12,000 years ago. Compared to this human antiquity, written history covers only a brief fraction of our species' lifetime. Although written history has clear limits in what it can tell us about the past, other sciences, including archaeology, paleontology, and genetics can help us learn more about our truly ancient history.

This book is not a comprehensive history of the human species but a collection of accounts illustrating the ways in which genetic data can be used to unravel our past. Some chapters deal with events that took place millions of years ago, whereas others focus on events that occurred within the past few hundred years. Some chapters focus on our entire species, whereas others deal with specific regions and populations. Various examples are used to show where and when genetic data can inform us about history, though there are several examples that illustrate that the answers are not always clear.

I have structured the remainder of this book along chronological lines, starting with events that took place millions of years ago, moving on to consider events over the past 150,000 years, and then to events over tens of thousands of years, thousands of years, and then, finally, recent events over the past few centuries. No attempt is made to systematically organize all the little bits and pieces of information about genetics and evolution, an approach typically used in a textbook. Instead, I jump right into questions about our history and explain the relevant genetics and evolutionary theory as needed.

Chapter 2 looks at our history from a very broad evolutionary perspective, specifically our relationship with our closest living relatives, the African apes. You may be familiar with the often-cited statement that more than 98 percent of our DNA is identical to that of chimpanzees (one of three species of African apes). What exactly does this number mean? When and where did the common ancestor live? When did we take one evolutionary path and the chimpanzee another? What changed in our line, and what changed in theirs? Although the study of relationships between species has long been a focus of anatomical analysis of living and fossil species, it can also be informed by genetic data, which reveal some answers to these questions.

Chapter 3 is concerned with the subject of modern human origins that I mentioned earlier. Much of the current controversy in anthropology focuses on events that took place over the past 150,000 years, a time characterized by the origin and spread of what are often referred to as "anatomically modern humans." To some, the fossil record reveals a clear pattern of a brand-new species *(Homo sapiens)* emerging in Africa and then dispersing across the planet, giving all of us a rather recent origin. Others interpret the fossil record differently, seeing evidence of a much more ancient web of geographically dispersed human populations interconnected by migration. Since the mid-1980s, genetic data have increasingly been used to address the questions of modern human origins. Did all humans over the past 2 million years belong to the same species? Where and when did modern humans appear? Much of the work over the past fifteen years or so has been interpreted as proof that our species has a recent origin in Africa and that human populations outside of Africa belonged to a different species that subsequently became extinct. Maybe so, but I will show that this case is not nearly as strong as some believe, and many questions remain about how to interpret genetic data.

Chapter 4 deals in more detail with this problem, looking specifically at one of those groups of early humans that are often claimed to have become extinct—the enigmatic Neandertals of Europe and the Middle East. Neandertals have had a long and bumpy ride throughout the history of anthropology. Although Neandertals were first thought to be a strange separate species, the steady accumulation of data throughout the twentieth century suggested that they were, in fact, simply a variant of ancient humans. Most anthropologists tended to consider them a different subspecies of human, *Homo sapiens neanderthalensis,* who were part of our ancestry. In recent years, this view has been challenged, and the idea that Neandertals were indeed a separate species *(Homo neanderthalensis)* has become accepted by a growing number of anthropologists. Part of this shift has to do with changing interpretations of fossils, but it also reflects some direct evidence of Neandertal genes. Since 1997, DNA has been extracted from several Neandertal fossils. Although the data are much less impressive than one might expect in an age conditioned by the wonders of *Jurassic Park,* they nonetheless provide a direct window on the past. As with other genetic data relating to the modern human origins question, the results at first seemed to support the view that Neandertals were a separate species, but as will be described later, doubts about this conclusion continue to linger.

Chapter 5 focuses on the relationships among geography, history, and the genetic variations we see in living humans. What can we learn about our past by examining global patterns of diversity in the world today? How much of what we see in the world today has been shaped by adaptation to different environments, and how much reflects the past history and geographic connections of different human populations? This chapter provides some background on the type of data and methods we use to reconstruct population history from the genetics of living humans. By examining global patterns of genetic diversity, this chapter provides an overview for the more detailed regional studies described in later chapters.

The next three chapters look at the history of groups in specific regions of the world, focusing primarily on group origins and relationships with groups in other regions. Chapter 6 considers the origin of Native Americans, the first humans to live in the New World. The genetic data agree with data obtained from archaeology and fossils; the first Americans came from Asia. However, the exact history is more complicated than that, and questions abound concerning the nature of this immigration, including when and how frequently it occurred and exactly where the immigrants came from. The question of Native American origins also ties into current arguments regarding the ancestry and custodianship of ancient human remains, including the case of the ancient skeleton dubbed "Kennewick Man."

Chapter 7 turns to the question of the origin of agriculture in Europe. Although agriculture appeared independently in many different parts of the world, the archaeological evidence supports the idea that agriculture spread into Europe from the Middle East. The question from an evolutionary perspective is whether the spread of this cultural innovation correlated with the spread of genes. Was agriculture spread culturally, from group to group, without any movement of genes? Or did agriculture arise because groups of farmers were moving into Europe? If so, to what extent did they replace, or mix with, preexisting European populations? In other words, what spread throughout Europe—farming or farmers?

Chapter 8 also deals with the question of geographic origins but focuses on a case where humans "recently" moved into a previously uninhabited region—the Pacific Islands. The human occupation of the Pacific Islands, with the intimidating barrier of vast stretches of ocean and the technology needed to cross them, has long been a favorite topic for anthropological investigation. Polynesian populations spread across the Pacific Ocean over the past several thousand years. Where did they come from, and how did they get there? Did these people originate in eastern Asia, as suggested by some archaeologists, and then spread rapidly into the Pacific Islands by themselves, or was there genetic mixing with other Pacific populations, such as the Melanesians, along the way?

Chapter 9 examines the relationship between genetics and history in a single nation—Ireland, the focus of much of my professional career. Here, I offer three separate tales about Irish genetic history. The first deals with the debate over the origin of a social group known as the Irish Travellers, a group often described as being culturally similar to Gypsy populations. Were the Travellers actually a Gypsy population that moved into Ireland, or were they a group of Irish people that adopted a similar lifestyle? The use of genetic data to test these ideas is explored. The second case study focuses on populations along the west coast of Ireland and anthropometric data suggesting that they have English roots. The third case study examines the population history of the entire island of Ireland, presenting evidence for the genetic impact of Viking invasion more than 1,200 years ago and the more recent impact of immigration from England and Wales.

The final chapter deals with genetic admixture, the mixing of gene pools of people that had previously been separated by time or distance. Historical events have often led to the geographic movement of people into different lands, with subsequent genetic admixture. This chapter examines three case studies of admixture. The first is the mixing of Native American and European genes in Mexican and Mexican American populations. How can we determine relative ancestral contributions? How have specific circumstances and population history affected rates of admixture? Similar questions are asked in the second case study, which focuses on the formation of African American gene pools. What do the genetic data tell us about the relative amounts of African and European ancestry in African Americans? Can we even make any statements about a single homogeneous African American gene pool, or is this an illusion based on our confusing concepts of cultural identity with those of genetic ancestry? Can genetic data be used to resolve specific historical questions, such as the long-standing controversy about whether Thomas Jefferson fathered any of Sally Hemings's children? The third case study in this chapter considers the genetic relationship between Jewish and non-Jewish populations in Europe, North Africa, and the Middle East. Is there any relationship between concepts of genetic ancestry and the culturally defined category "Jewish"? Is there any evidence of mixing between European Jewish populations and their non-Jewish neighbors? Finally, this chapter closes with an examination of the relationship between genetic ancestry and cultural identity.

There are many other possible topics of interest when considering the broad picture of genetics and history. My attempts here are designed to give some specific examples of how anthropologists and geneticists use genetic data to explore human history. Some of the examples I have chosen are very specific historically or geographically, and others are broader in focus. Although your own recent history may or may not be reflected in some of the later chapters, the early ones are broad in scope and pertain to all of us, regardless of our specific group affinity or recent history.

One of the reasons that anthropology appeals to me is its focus on the *broad* picture, features that we all share as humans. Although each group has its own history, there are certain elements of a common history that are universal and include *all* of us.

Our broad history, dealing with the very beginning of the line leading to present-day humanity, is shared by all humans, regardless of ancestry, in all parts of the world. It all started six million years ago . . .

TWO

The Naked Ape

Which of the following is the most different—an orange, an apple, a pear, or a potato? You've probably taken many similar tests that ask you to figure out which of a set of items does not belong. The correct answer here would be potato because all of the other items are fruits. Sometimes the questions are a bit trickier. One of my favorite examples comes from a self-scoring IQ test I bought years ago. The question shows drawings of five objects—a saw, a knife, a spoon, a shovel, and a screwdriver—and the goal is to figure out which of these five objects is least like the other four. My answer was the shovel, because I reasoned that each of the other four objects was held in one hand, whereas the shovel required two hands. Logical, but incorrect according to the answer sheet, which stated that the knife was the odd item, since it began with the letter *k* and all of the other objects began with the letter *s*.

If you think about it, both answers were correct depending on the criterion you use to categorize the objects. Classification of objects can be done in a variety of ways and for different purposes. The one thing that helps in the above example is that you know there are supposed to be only two groups: one group consisting of four objects with some characteristic in common, and another group consisting of a single odd object. Suppose you were given the same list and you were instructed to arrange the objects into groups, but you were not told the number of groups. You might lump them all into one group, labeled "tools." Alternatively, you might classify them as I did, into objects requiring one hand versus two, or into groups according to the first letter of each object. You could even arrange them into three different groups according to the number of syllables in each: you would have one group of objects with a single syllable (saw, knife, spoon), another group with objects with two syllables (shovel), and a third group with objects with three syllables (screwdriver).

Classification is fundamental to understanding the variation and evolution of life. Consider the following list of animals: bat, mouse, canary, whale, and goldfish. What are the different ways these animals could be organized into different groups? The number of groups, and the specific animals within each, depends upon the criterion you use to create the classification. Suppose, for example, you use body size as a trait for sorting these animals. You would wind up with two groups, one with small animals (bat, mouse, canary, and goldfish), and the second with large animals (whale). An alternative classification can be made based on how these animals move about, in which case you would come up with one group consisting of animals that fly (bat and canary), a second group with those that move about on land (mouse), and a third group with those that swim in water (whale and goldfish).

Which is correct? Again, it depends on the purpose of the classification. As you read the list of animals, you were probably thinking that there are three groups: mammals (bat, mouse, and whale), birds (canary), and fish (goldfish). If this was your answer, it probably came as a result of information you had previously learned regarding the classification of living organisms, specifically the division of vertebrates into groups known as mammals, birds, and fish (as well as amphibians and reptiles, which are not represented in this particular example). This classification scheme is usually first learned in grade school and then revisited in a number of biology classes in middle school, high school, and college. Most everyone is familiar with this classification system, and therefore, if I gave you a list of five organisms—say, bat, mouse, sparrow, whale, and horse—and asked you to identify the animal least like the other four, you would most likely pick the sparrow because it is a bird and the other four are all mammals.

The "correct" answer is chosen because of our knowledge of how organisms are classified in biological science. I recall spending a great deal of time in high school biology learning the various groups and subgroups of living organisms. Given constant reinforcement, it is natural for us to view the world in this way. Less attention is often given to *how* these classifications arose and *why* they are important in understanding biology.

Philosophical debates over classification take on special significance when we turn to the question, "What are humans?" This question forms a major focus of much human effort, ranging from the sciences to religion to philosophy. This question is put to God in Psalms 8:4–6:

> What is man, that thou art mindful of him? and the son of man, that thou visitest him? For thou hast made him a little lower than the angels, and hast crowned him with glory and honor. Thou madest him to have dominion over the works of thy hands; thou hast put all things under his feet.

This passage reflects a common view of humanity as having been created to be special among living creatures but still "lower than the angels." A view of humans as special and superior is found in much of Western philosophy and is perhaps best summarized by Aristotle's "Chain of Being," which ranked humans at the top of a "scale of nature."[1]

The question, "What are humans?" concerns the extent to which we are part of, rather than separate from, the rest of the animal kingdom. What is the best way to classify humans? Should we focus on our spiritual natures, placing us lower than the angels, or should we be considered as simply another species? Are we better described as "featherless bipeds" or "naked apes"? Are the differences between us and other organisms qualitative or quantitative? Our culture holds both views. On the one hand, we all realize that humans are animals in a zoological sense, yet we frequently tell our children not to act like animals, showing how we use the term "animal" in different ways. Judeo-Christian tradition suggests that humans have special status in the mind of a creator and hold dominion over all other creatures because *we* were created in God's image, not the apes. Yet, studies of genetics show us that we are more than 98 percent genetically similar to African apes.

Our Place in Nature

This chapter examines our relationship to other organisms from a biological perspective. What is our place in nature? What can genetics tell us about the pattern and degree of this relationship? Who are our closest living relatives, and how long have we been on a separate evolutionary path from them?

The science of taxonomy, the classification of living creatures, has its roots in the work of Carolus Linnaeus, an eighteenth-century Swedish naturalist who developed the system of hierarchical classification you probably remember from high school. It starts out with a broad category, the kingdom. Each kingdom can be subdivided into a number of different phyla, each phylum can be subdivided into a number of different classes, each class can be subdivided into a number of different orders, and so on. Organisms share certain characteristics with all other organisms at the same level in the hierarchy. Humans, for example, belong to the animal kingdom and thus share certain characteristics, such as mobility and a need for ingested food, with all other animals, ranging from insects to apes.

A "traditional" classification of humans is given in Table 2.1. We belong to the animal kingdom and specifically to the subphylum of vertebrates, defined by the presence of a segmented spinal column. There are five classes of vertebrates: fish, amphibians, reptiles, birds, and mammals. We are mammals, defined by the presence of mammary glands for breast-feeding and the ability to maintain a constant body temperature, among other characteristics. There are three different subclasses of mammals in the world today: egg-laying mammals, such as the duck-billed platypus; marsupial mammals, such as the kangaroo and opossum; and the placental mammals, the group to which we belong. Within the subclass of placental mammals, humans belong to the order primates, defined as having forward-facing eyes, grasping ability in the hands, nails rather than claws, and a number of other anatomical characteristics. Within the primates we belong to the suborder anthropoids, a group composed of monkeys, apes, and humans. There are two major groups of anthropoids: the platyrrhines (New World monkeys) and the catarrhines (Old World monkeys, apes, and humans), distinguished primarily by nostril shape and orientation.

Among the Old World anthropoids, we are hominoids, a group made up of the living apes and ourselves. Hominoids share a number of traits. They all lack a tail, a trait found in all other primates, and all have similar cusp patterns on their molar teeth. Hominoids also share aspects of shoulder anatomy, allowing them to raise their arms above their heads easily. Although there were many different species of hominoids between 10 and 20 million years ago, there are only a handful of species alive today. These have usually been placed in one of three different zoological families (Figure 2.1), although we shall see that this traditional classification has been increasingly challenged in recent years.

The family known as hylobatids, or lesser apes, consists of the gibbon, a small Asian ape capable of fantastic aerial acrobatics. Pongids, also known as the great apes, are the second zoological family. There are four different species of great ape alive today: one species of Asian great ape and three species of African great apes. The Asian great ape—the orangutan—is a

Table 2.1 Traditional Classification of Humans

Taxonomic Level	Placement of Humans	Some characteristics[1]
Kingdom	Animals	Capable of moving; relies on eating other organisms
Phylum	Chordates	Has a spinal cord
Subphylum	Vertebrates	Has a backbone
Class	Mammals	Has mammary glands
Subclass	Placental mammals	Has a placenta during pregnancy
Order	Primates	Grasping hands and depth perception
Suborder	Anthropoids	Larger bodies and brains; less primitive
Infraorder	Catarrhines	Old World primates with particular orientation of nostrils
Superfamily	Hominoids	Lacks a tail; shoulder structure adapted for climbing
Family	Hominids	Bipedal
Genus	*Homo*	Large brain; reliant on tools
Species	*sapiens*	

[1]These are illustrative, not comprehensive. Each group has other defining characteristics.

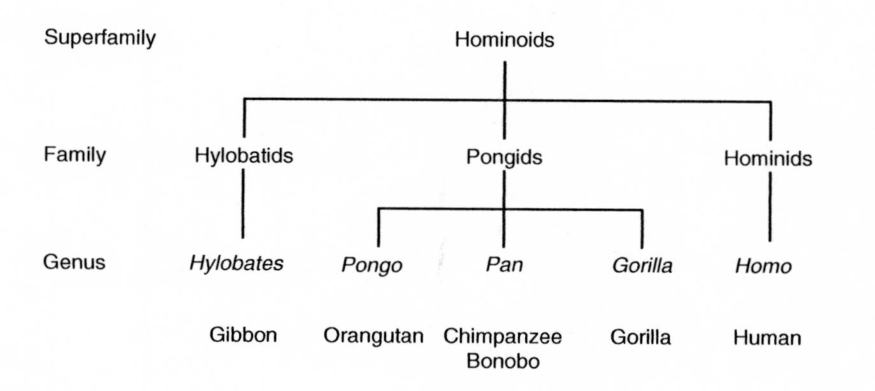

Figure 2.1 Traditional classification of the hominoids into three zoological families: hylobatids (lesser apes), pongids (great apes), and hominids (humans). This classification is falling out of favor as a result of finding that humans and African apes are more closely related than either is to orangutans or gibbons.

large-bodied ape with reddish-brown fur. Orangutans tend to live in small social groups consisting of a mother and her dependent offspring. In the trees, they are agile climbers, but they walk on all fours when on the ground, balling their front hands into fists for support when moving. As with all apes, orangutan arms are longer then their legs.

There are three species of African apes: gorillas, chimpanzees, and bonobos. All three have black fur and rest on their knuckles when walking on the ground. Gorillas are the largest African apes; they live in small social groups of a single adult male, several adult females, and their offspring. Gorillas rely extensively on a vegetarian diet. Chimpanzees are smaller and live in larger social groups made up of a number of adult males and females and their offspring. The lesser-known bonobo is physically very similar to the chimpanzee but has a more slender body build (bonobos were once known by the misleading label "pygmy chimpanzee," although they are not smaller than chimpanzees and are now known to be a distinct species). Bonobos are different behaviorally from chimps in several ways; females are likely to be as dominant (or more dominant) than males, and they engage in a great deal of sex play for resolving tension in the group.

Finally, there is the third family: the hominids, to which humans belong. This zoological family contains only one living species—ourselves—but the term "hominid" is also used to describe fossil ancestors that share our bipedal stance. Although we share many anatomical features with the great apes, it is obvious that we are also different in many ways.

Who Is Our Closest Living Relative?

It has long been known that humans are more closely related to the great apes and less closely related to the lesser apes, the gibbons. What has changed over the past century is our view of relatedness between humans and the great apes, a change that has been strongly influenced by the application of genetic data to the question of relationship between species.

We can look at this problem by comparing humans to the great apes. Consider these five species: orangutan, gorilla, chimpanzee, bonobo, and humans. Which of these are most closely related to each other? Which are the most different? You've probably seen apes in a zoo or on television, and the answer seems pretty obvious; humans look the most distinct. All of the great apes share characteristics not found in humans. Many of these differences are easy to see with only a casual glance. The great apes are hairy, and we are relatively hairless (actually, our hairs are smaller and finer, but this does give us the appearance of a "naked ape"). All of the great apes have arms that are longer than their legs, whereas humans have the reverse—our legs are longer than our arms. The great apes all walk on four limbs (although the orangutan does this somewhat differently than the others), whereas we walk on two limbs. Our brains are much larger relative to body size, and our faces and teeth are smaller. Our canine teeth are small, compared to the larger and projecting canine teeth of the great apes.

We can add behavioral differences to this list of comparisons. Humans use a symbolic language. Although some great apes have been taught the rudiments of American Sign Language and other symbolic languages, their abilities in terms of both vocabulary and grammar are quite different from ours. Although an ASL-trained chimp can request an apple or ask to be tickled, no ape discusses world events, the meaning of life, or what they thought about a movie. Humans make and use much more complex technology than apes. Some ape species have invented simple forms of tools, such as sticks for fishing termites out of mounds; nevertheless, the technological and cultural achievements of humans are clearly quite distinct. Even though studies of ape behavior have shown some ability for language acquisition, technology, and culture, suggesting that the differences between ape and human behavior are more a matter of degree than of uniqueness, the gap is still quite dramatic.

Studies of primate anatomy and behavior portray our relationship to the great apes. They are similar to us in some ways but quite different in others. The overall level of similarity, viewed from the lens of superficial anatomy and behavior, suggests that the great apes, as a group, are closely related to us, but that they are more closely related to each other. Thus, if asked to consider the five hominoid species and to pick the one that is most different, most people would undoubtedly choose the human—the bipedal, big-brained naked ape with the small canine teeth. The great apes all look more similar to each other than to us.

Given these overall dissimilarities, it comes as no surprise that scientists dating back to Linnaeus placed humans in a zoological family separate from the great apes. We are hominids, and the great apes are pongids. Our placement in the same superfamily (hominoids) but in a separate family (hominids) seems to reflect our overall view of our place in nature. We are part of the animal world, but we are also unique. I think that placing ourselves in a separate zoological family is reassuring to some extent. We realize that we are close to the apes, but we don't want to be too close!

Is this traditional classification "correct"? As I argued earlier, whether a classification is correct depends largely on the purpose of the classification. If our goal is to order organisms by their level of *overall* similarity, physically or behaviorally, the traditional classification of pongids and hominids works fine. Humans *do* look different from the great apes, and consequently it seems reasonable to place them in a separate group. However, genetic comparisons of the great apes and humans provide a different picture of our evolutionary relationships.

Genetic Comparisons of Apes and Humans

The use of genetic data to assess relationships between different primate species began in 1904 when George Nutall suggested that analysis of blood chemistry could potentially tell us something about our relationship to other primates. Nutall looked at immunological reactions between the blood of different species in order to evaluate their relationship to each other. His results showed that humans were more similar to apes than to monkeys and that apes and humans were more closely related to Old World monkeys than to New World monkeys.[2]

Results that are more precise came in the early 1960s, when Morris Goodman compared the immunology of different primate species by examining the reaction of proteins in the serum of blood to antibodies from other species. Goodman viewed these reactions as a measure of genetic difference between species, which could be used to make statements regarding overall genetic relationship. His major finding was that humans and the African apes are more closely related to each other than either is to the Asian great ape, the orangutan. These results further suggested that humans and African apes shared a more recent common ancestor with each other than with the orangutan. Based on these results, Goodman suggested that the traditional classification of pongids and hominids was inappropriate and that humans and African apes should be placed in the same zoological family.

There was great resistance to the notion of changing the classification, however. The eminent biologist George Gaylord Simpson, for example, argued that human adaptations were so unique, relative to those of apes, that they should not be placed in the same zoological family.[3] The debate over whether to use evolutionary relationships or adaptations as the basis for forming classifications continues to this day.

The immunological research of Goodman and others in the 1960s led to the development of a new field—molecular anthropology—which examines the evolutionary relationships of different primate species (particularly humans) based on comparative biochemistry at the molecular level. One of the new leaders of this field was Vincent Sarich, then a graduate student at the University of California at Berkeley. Sarich originally enrolled as a graduate student in chemistry but soon turned his attention and laboratory skills to the field of anthropology.[4] Working with Allan Wilson, a Berkeley professor of chemistry, Sarich focused his initial attention on the albumin protein as a means of inferring genetic relationships among primate species.

Serum albumin was taken from a number of different primate species. The albumin for a given species was then injected into rabbits, whose immune system recognized the invading molecules and produced antibodies, since primate albumin is quite different from rabbit albumin. These antibodies were then mixed with the albumin from other primate species to measure the degree of reactivity. The closer two primate species are genetically, the greater the reaction. This procedure was then repeated using different primate species. The range of reactions was then converted to a series of "immunological distances" (ID) that represent the biochemical differences between species. Species that are genetically and biochemically similar to each other have a low ID, reflecting greater reactivity, whereas species that are more dissimilar have a higher ID. For example, Sarich found that the difference between humans and chimps was 7 units, compared to a distance of 32 units between humans and rhesus monkeys and a distance of 30 units between chimpanzees and rhesus monkeys. These numbers show that humans and chimpanzees are more closely related to each other than either is to the rhesus monkey, an Old World monkey.[5] Immunological comparisons were eventually done on several proteins, including albumin and transferrins. The results confirmed Goodman's conclusion; humans and African apes were more closely related to each other than either was related to the orangutan (Figure 2.2).

Over time, the ability to make genetic comparisons increased, first with the sequencing of amino acids, chemical units that make up proteins, and later with the direct sequencing of the genetic code—the DNA molecule.[6] Comparison of DNA sequences provides the ultimate in genetic analysis.

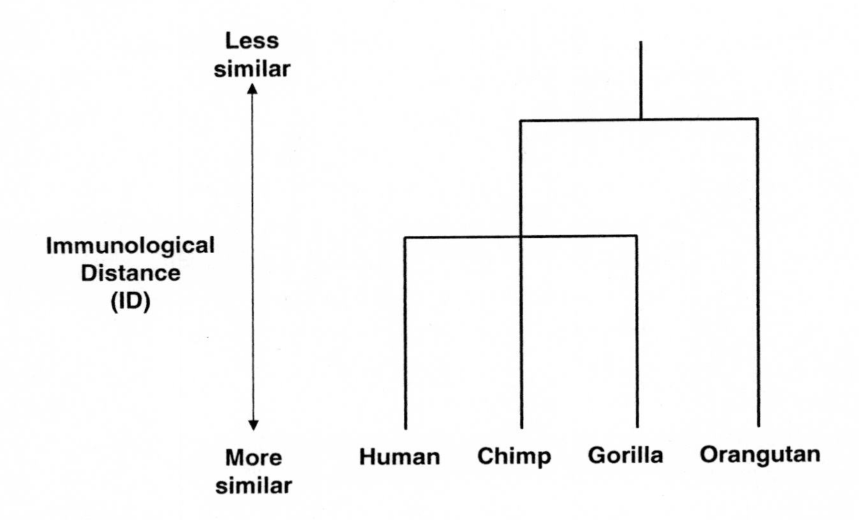

Figure 2.2 Relationships among great apes and humans based on immunological distance showing how humans and African apes are more closely related to each other than either is to the orangutan. This pattern of similarity supports the view that the orangutan line split off first from the rest of the hominoids. Subsequent analyses have demonstrated that humans and chimps are even more closely related to each other than either is to the gorilla.

DNA is a long molecule made up of four different chemical units, known as bases. The four bases are A (adenine), T (thymine), G (guanine), and C (cytosine). The genetic "code" of DNA consists of sequences of these bases, such as AATCGCGT or CCTCCGATTTAACG. Much of an organism's genetic code is found in long strands (chromosomes) found in pairs within the nucleus of cells. Some sections of DNA are codes for specific protein sequences, and these sections are usually labeled as "genes." Much of the DNA, however, is not code for anything and has no apparent purpose, but it is still very useful for determining evolutionary relationships. Some use the term "gene" to refer to any DNA sequence, coding or not, a convention I will use throughout this book to simplify description.

The codes for amino acids consist of sequences of three bases. For example, the DNA sequence AAA codes for the amino acid phenylalanine, whereas the sequence GTG codes for the amino acid histidine. Some DNA serves as "punctuation" in the genetic message; for example, the sequence ACT terminates a string of amino acids. The basic rule is that DNA codes for amino acids, which are the building blocks of proteins, which in turn make up various tissues and control development and other biological

Table 2.2 The Genetic Code

Code	Amino acid	Code	Amino acid	Code	Amino acid	Code	Amino acid
AAA	Phenylalanine	ATA	Tyrosine	ACA	Cysteine	AGA	Serine
AAT	Leucine	ATT	Stop	ACT	Stop	AGT	Serine
AAC	Leucine	ATC	Stop	ACC	Tryptophan	AGC	Serine
AAG	Phenylalanine	ATG	Tyrosine	ACG	Cysteine	AGG	Serine
TAA	Isoleucine	TTA	Asparagine	TCA	Serine	TGA	Threonine
TAT	Isoleucine	TTT	Lysine	TCT	Arginine	TGT	Threonine
TAC	Methionine	TTC	Lysine	TCC	Arginine	TGC	Threonine
TAG	Isoleucine	TTG	Asparagine	TCG	Serine	TGG	Threonine
CAA	Valine	CTA	Aspartic acid	CCA	Glycine	CGA	Alanine
CAT	Valine	CTT	Glutamic acid	CCT	Glycine	CGT	Alanine
CAC	Valine	CTC	Glutamic acid	CCC	Glycine	CGC	Alanine
CAG	Valine	CTG	Aspartic acid	CCG	Glycine	CGG	Alanine
GAA	Leucine	GTA	Histidine	GCA	Arginine	GGA	Proline
GAT	Leucine	GTT	Glutamine	GCT	Arginine	GGT	Proline
GAC	Leucine	GTC	Glutamine	GCC	Arginine	GGC	Proline
GAG	Leucine	GTG	Histidine	GCG	Arginine	GGG	Proline

There are 64 different combinations of the four bases (A, T, C, G) taken in three-base units. These 64 code for 20 amino acids and a "stop" message. Many amino acids can be specified by slightly different sequences.

phenomenon. Table 2.2 lists all of the three-base combinations for DNA and their associated amino acids (or termination message). Note that some amino acids can be specified by a number of different sequences, thus providing some redundancy.

DNA is passed on, generation to generation, over time. Although the DNA is usually copied correctly, there are occasional "errors" resulting in a change in the underlying genetic code. Such changes are known as mutations. As mutations occur independently in different species, those species become increasingly dissimilar. By looking at amino acid sequences, which in turn reflect underlying differences in the genetic code, we can obtain a relative idea of how closely related two (or more) species are to each other. Figure 2.3 shows an example comparing five primate species for the amino acid sequences of one of the protein chains that make up hemoglobin, the molecule that carries oxygen. This figure shows only those amino acid positions where there is at least one difference among the five species; positions where all five species share the same amino acid are not shown, since they don't contribute anything to an analysis of evolutionary change.[7]

Several comparisons can be made. First, humans and chimpanzees have all of the same amino acids for this molecule. Second, gorillas differ from

Amino acid position	Human	Chimpanzee	Gorilla	Gibbon	Rhesus monkey
9	Serine	Serine	Serine	Serine	Asparagine
13	Alanine	Alanine	Alanine	Alanine	Threonine
33	Valine	Valine	Valine	Valine	Leucine
50	Threonine	Threonine	Threonine	Threonine	Serine
76	Alanine	Alanine	Alanine	Alanine	Asparagine
80	Alanine	Alanine	Alanine	Aspartic acid	Alanine
87	Threonine	Threonine	Threonine	Lysine	Glutamine
104	Arginine	Arginine	Lysine	Arginine	Lysine
125	Proline	Proline	Proline	Glutamine	Glutamine

Figure 2.3 Comparison of amino acid sequences for one of the protein chains of the hemoglobin molecule. Only those amino acid positions that show at least one difference among the five species are shown. Differences from humans are shaded in gray. Note that humans are most similar to chimpanzees, followed by the gorilla, the gibbon, and then the rhesus monkey. Source: Weiss and Mann (1990).

humans and chimps in only one amino acid; they have the amino acid lysine at position 104, whereas humans and chimps have the amino acid arginine. Gibbon hemoglobin is more distinct, with three differences from humans, and the rhesus monkey (an Old World monkey) is the most different, with eight differences. Just looking at the number of amino acid differences, we see that humans are very closely related to the African apes, particularly the chimpanzee.

Comparison of amino acid sequences gets a bit more complicated because the same amino acid can result from different DNA sequences. For example, both the sequence TTT and the sequence TTC code for the amino acid lysine. Some amino acids, such as glycine, have even more possible codes (CCA, CCT, CCC, CCG), whereas some have only one, such as tryptophan (ACC). When we compare two species and see different amino acids at the same position in a protein chain, we don't know specifically how many mutations have taken place. Another problem is that some mutations are "silent," meaning that they don't show up in a comparison of amino acids. For example, imagine a mutation in the third base of the DNA sequence GGA, where A mutates into T, giving the sequence

GGT. This mutation would not be picked up by amino acid analysis since both GGA and GGT code for the same amino acid—proline. Fortunately, methods have been developed to get around some of these problems.

A number of studies of amino acid sequences in primates show the same basic pattern: Humans and African apes are more closely related to each other than either is to the orangutan. Once again, the genetic evidence conflicts with the traditional division of the hominoids into three families of humans, great apes, and lesser apes. In addition to confirming our kinship with the African apes, amino acid sequence analysis has also shown how closely we are related. A classic study by Mary-Claire King and Allan Wilson in 1975 showed that humans and chimpanzees share over 98 percent of their DNA, a finding later replicated by other genetic methods.[8] By itself, this number does not have much use, since it is necessary to compare both humans and apes to other species. However, the high correspondence did serve to reinforce a changing view of the world; humans and apes may not be that dissimilar after all.

Over the past fifteen years or so, major advances in genetic technology have allowed the direct comparison of DNA sequences, thus getting around some of the problems associated with amino acid analysis. The simplest way to compare DNA sequences is simply to count the number of differences between two sequences. For example, given the following two sequences,

Sequence #1: GGT**G**CAATGGTT**A**CGC
Sequence #2: GGT**C**CAATGGTT**T**CGC

we see that there are two differences, one in the fourth position (G versus C) and one in the thirteenth position (A versus T). When comparing a number of sequences, either from individuals or from species, each sequence is compared to all other sequences in the study so that all possible pairs are compared. For example, if we had DNA sequences from four species, labeled 1 through 4, we would compare 1 with 2, 1 with 3, 1 with 4, 2 with 3, 2 with 4, and 3 with 4. In each case, we would count the number of differences in the bases (see Figure 2.4 for an example).

One of the most powerful genetic tools developed in the past two decades has been the analysis of mitochondrial DNA sequences. Most of our DNA is contained in the cell nucleus (and called nuclear DNA). A small amount of DNA, a little over 16,000 bases in length, is contained in

Comparing DNA Sequences

Here are the sequences:

Sequence 1:	G G C T T A C C G G T A C C G
Sequence 2:	G G G T T A C C G G T A C C G
Sequence 3:	G G C T A A G C G G T A C C G
Sequence 4:	G G C T A A G C G G A A C C G

Here are the number of differences between each pair of sequences:

Sequences 1 and 2:	1 difference
Sequences 1 and 3:	2 differences
Sequences 1 and 4:	3 differences
Sequences 2 and 3:	3 differences
Sequences 2 and 4:	4 differences
Sequences 3 and 4:	1 difference

Figure 2.4 A hypothetical example comparing DNA sequences. To get an idea of how dissimilar species are, one simply counts the number of differences between all possible pairs of sequences.

the mitochondria, the numerous organelles responsible for energy production in the cell. Mitochondrial DNA (abbreviated as "mtDNA") differs from nuclear DNA in its pattern of inheritance; nuclear DNA is inherited from both parents, whereas mitochondrial DNA is inherited from only the mother.[9]

Human nuclear DNA is contained in 23 pairs of chromosomes inside the nucleus of body cells. In general, you have inherited one of each pair from your mother and one of each pair from your father. Inheritance actually is a bit more complicated because of recombination. During the production of sex cells, which contain one of each chromosome pair, pieces of the maternal and paternal chromosome can swap with each other, producing a situation where a given chromosome in your sex cell might contain mostly one parent's DNA with a smaller amount of DNA from the other parent. The mixing of parental chromosomes in each generation, combined with recombination, generates a great deal of genetic variation, but it makes tracing ancestry difficult.

At conception, the nuclear DNA in the sperm (containing 23 chromosomes) combines with the nuclear DNA in the egg (containing 23 chromosomes) to produce 23 *pairs* of chromosomes in the zygote, the fertilized egg. Mitochondrial DNA is not inherited in this way. Instead, the mitochondria comes from the mother's egg cell, and hence so does the mitochondrial DNA. Your mitochondrial DNA was inherited from your mother, who inherited it from her mother, who inherited it from her mother, and so on, back in time. Your father does not contribute to your mitochondrial DNA.

This maternal inheritance makes tracing ancestry easier because there is no mixing of genetic material. It also means you have fewer mitochondrial ancestors in the past. For nuclear DNA, the maximum number of ancestors doubles each generation in the past. You have two ancestors a generation ago—your parents. You have four ancestors two generations in the past—your grandparents. Likewise, you have a maximum of eight great-grandparents, sixteen great-great-grandparents, and so on. The situation is different for mitochondrial DNA. You have only one ancestor a generation ago—your mother. You also have only one ancestor two generations ago—your mother's mother. For each generation in the past you have only one mitochondrial DNA ancestor.

A good analogy for the inheritance of mitochondrial DNA is the inheritance of last names in many Western societies. Although there are exceptions, the usual pattern in the past few centuries has been for a child to take his or her father's name. Male children keep this name their entire life and pass it on to their children. In this case, surnames are "inherited" through the father's line. Although I have four grandparents, only one (my father's father) was born with the surname Relethford. Likewise, I carry the mitochondrial DNA of my mother's mother.

If we compare the mitochondrial sequences of different individuals or species, we get a measure of how similar they are. If they are different, it is because mutations have occurred over time. The fewer the mutations, the smaller the genetic difference likely to have accumulated over time. Comparison of mitochondrial DNA sequences is a useful means of examining genetic relationships between different species. In the case of the hominoids, the results are clear and consistent: Humans and African apes are more similar to each other than either is to the orangutan (see Figure 2.5 for an example).[10] Such studies also show chimpanzees and bonobos to be very closely related, as expected from their similar anatomy. They also suggest that chimpanzees and bonobos are both more similar to humans than either is to the gorilla. Viewed from this perspective, humans fall clearly

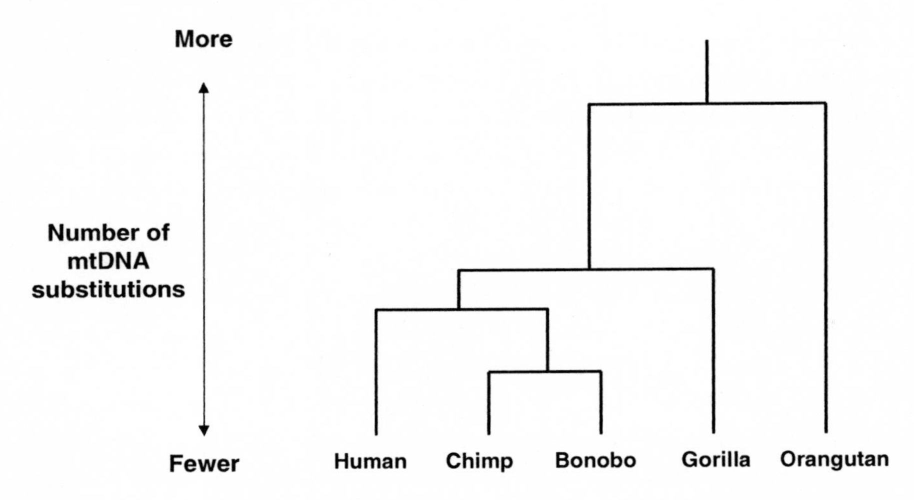

Figure 2.5 Evolutionary relationships between humans and great apes based on the number of mitochondrial DNA substitutions. Humans are most similar to chimps and bonobos, then gorillas, and then orangutans.
Source: Horai et al. (1995).

within the group of African apes. Perhaps the label "naked ape," assigned by zoologist Desmond Morris, is an apt description after all.[11]

Dating the Split

These genetic studies have called into question the traditional grouping of humans and great apes as separate zoological families for the simple reason that we are more closely related to *some* of the great apes than others. These findings have major implications for debates about the rationale and use of different methods of classification, as well as for the evolutionary history of humans.

Results such as those shown in Figure 2.5 can be interpreted in terms of evolutionary history using the principle that genetic similarity is a measure of evolutionary kinship; the closer two species are genetically, the more recently they shared a common ancestor. Humans and great apes are all considered to be part of an evolutionary group that, at one time, had a common ancestor. What happened then? Using the genetic relationships as a guide, anthropologists suggest that the first branch in the family tree split between an Asian line, leading to modern day orangutans, and an African line, leading to the African apes and humans (which makes sense geographically, since the first hominid fossils are found in Africa). The African line then later split again—first between the line leading to the gorilla and that leading to chimps, bonobos, and humans, and then again between the human line and that leading to chimps and bonobos, and then yet again between the chimp line and the bonobo line. This reconstruction is based on the principle that the more similar two living species are genetically, the more recently they both split from a common ancestor. Since the orangutan is the least similar, its evolutionary line must have split off first. Thus, based on the genetic relationships shown in Figure 2.5, we can hypothesize a family tree, shown in Figure 2.6.

We wind up with a tree showing the evolutionary history of the great apes and humans. However, this tree reflects past events involving only those species that are alive today. It does not tell us anything about species that lived in the past but became extinct. If another line split off from the orangutan line later in time and then became extinct, we would not be able to detect it from genetics, because we have no living descendants. Thus, our genetic family tree is really only a piece of the larger picture, showing us the evolutionary relationships among only those species that have survived to the present day.

Also, the family tree doesn't tell us anything about what the various common ancestors looked like or where or when they lived. We can make logical inferences based on comparisons among the living species, but ultimately we need to go to the fossil record to see what species existed and where they might fit in this family tree. This is more complicated than it sounds, because a fossil ape might not lie directly on one of the branches of our tree; it might represent one of the extinct dead ends that we can't identify from genetic analysis.

Before the field of molecular anthropology took off, most of what we knew about the evolution of apes came from the fossil record. By the mid-1960s, we had accumulated fossil evidence of ape species living during the Miocene epoch (23 to 5 million years ago) in Africa, Europe, and Asia. Some of these fossil apes appeared to be not direct ancestors of the living great apes or humans but branches of the tree that had become extinct. Other fossil apes, however, seemed to be very similar in some anatomical features to living apes, particularly specimens assigned to the genus *Proconsul* that lived in Africa roughly 20 million years ago. Most of the remains at that time consisted of teeth and jaws, not unusual since these are the hardest parts of the skeleton and are more likely to be preserved

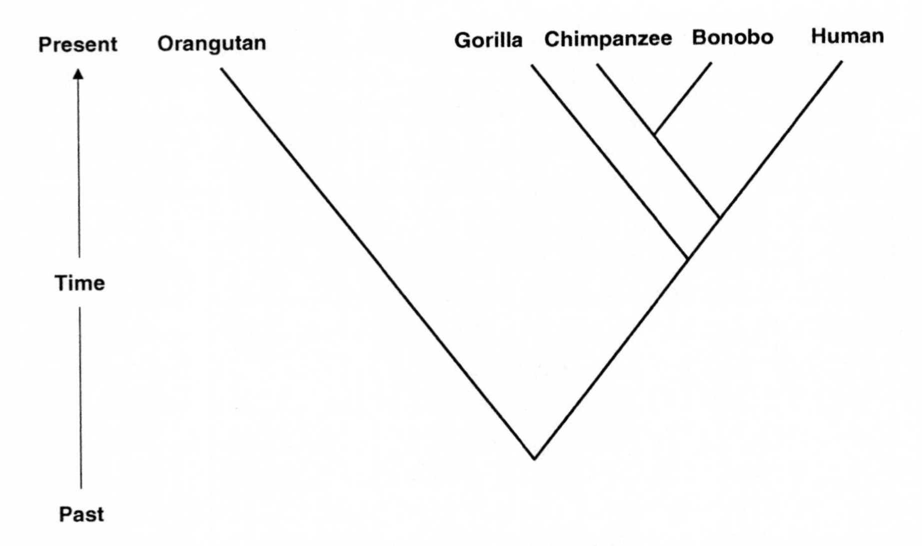

Figure 2.6 Genetic dissimilarity between species as a function of time. The more time that has elapsed from a common ancestor, the more genetically dissimilar the two species. This picture suggests that orangutans branched off first from a common ancestor of great apes and humans. The gorilla line branched off later, followed by the split between the chimpanzee-bonobo line and the human line.

as fossils compared to small or delicate bones. The teeth were definitely ape-like, right down to the level of the cusp patterns on the lower back teeth. Furthermore, there was evidence of larger species and smaller species. Overall, the small ones looked a lot like chimpanzees, while the large ones looked a lot like gorillas. Perhaps these fossils represented the first African apes.

What about the first hominids—our ancestors? Some looked to dental remains of a fossil form then known as *Ramapithecus* (which has since been reclassified as a species in the genus *Sivapithecus*). Although very fragmentary, the *Ramapithecus* remains suggested some hominid affinity. First, the canine teeth were small, much like ours but unlike those of most living apes. Second, the jaw fragments, although broken, seemed to have a parabolic shape, again like humans but unlike living apes (whose jaws are more U-shaped with parallel rows of back teeth). There were other features as well that suggested *Ramapithecus* was our ancestor. Note, however, that there was no evidence of upright walking, considered by many to be a primary characteristic of hominids. The oldest specimens were found in Africa (named *Kenyapithecus* by some) and were 14 million years old. If this 14-million-year-old species was indeed a hominid, then the hominid line must have split off from the African ape line earlier in time, most likely in the range of 15 to 20 million years ago.

Not everyone agreed with this conclusion. A number of anthropologists questioned the hominid status of *Ramapithecus,* arguing that it was actually an ape or that we knew too little about its overall anatomy to say much of anything. Meanwhile, the development of molecular anthropology brought some new insights by using genetic data to estimate the date of the African ape-human split. This accomplishment began with the application of "molecular clocks" to the study of human evolution by Vincent Sarich and Allan Wilson.[12]

We have seen that genetic relationships of living species can be used to make evolutionary inferences about the past. Two species that are more genetically similar share a more recent common ancestor than other less genetically similar species. Given certain assumptions, measures of genetic similarity can be taken as proportional to the length of time since species shared a common ancestor. Imagine, for example, that you are comparing the DNA sequences of three species—A, B, and C. You find that species A and B are more similar genetically to each other than either is to species C. You would then conclude that species A and B shared a common ancestor more recently than with species C. Suppose we quantify these differences in similarity with some measure of genetic distance, where the higher the number, the more dissimilar the two species. There are many methods of computing the genetic distance between species. For now, though, let us simply imagine that our analysis gives us the following results (in genetic distance units):

The distance between species A and species B = 4
The distance between species A and species C = 12
The distance between species B and species C = 12

It is clear that A and B are more closely related to each other than to C. The distances between the three species is easily made into a tree showing the evolutionary history of the three species (Figure 2.7a).

What else can we say other than species C split off earlier in time? Given certain assumptions, we can state that the date at which species A and species B split from a common ancestor was one-third the date at which species C split from a common ancestor. This is done simply by noting that the distance between A and B is one-third that of A and C (or B and C), given the distances above ($4/12 = 1/3$). This still doesn't give us a date. However, if we know the actual date for at least one of the common ancestors in a tree, we can derive the rest. For example, imagine that we have a good fossil record on the initial split between species C and the common ancestor for A and B, and we know that this took place 9 million years ago. We can then derive the date at which A and B split off by multiplying 9 million years by $1/3$ to arrive at 3 million years (see Figure 2.7b).

Sarich and Wilson did just this using their albumin data. They looked at immunological distances between humans and a number of primate and other mammal species to generate a tree of evolutionary relationships. They then calibrated this tree by using known dates of mammalian evolution to estimate that approximately 1.67 units of immunological distance occur per million years of evolution. Given that 7 units of immunological distance separates humans and African apes, the split between the hominid and African ape lines occurred $7/1.67 = 4.2$ million years ago.[13] Using a number of calibration points and different methods of analysis, they consistently arrived at a date of 4 to 5 million years ago for the time of the split.[14]

This date was much later than the 15 to 20 million-year date suggested by some paleoanthropologists. For those who argued *Ramapithecus* was *not* a hominid, this wasn't a big deal. However, to those who considered *Ramapithecus* a hominid that lived *after* the hominid-ape split, these numbers were clearly wrong! Sarich and Wilson felt otherwise and debate ensued, perhaps made even more contentious by Sarich's statement that the molecular data on humans and African apes showed that "one no longer has the option of considering a fossil specimen older than about eight million years a hominid *no matter what it looks like*" (italics in original).[15] This statement did not sit well with a number of paleoanthropologists, since it implied rather forcefully that, when in doubt, one should ignore the fossils.

Thus began a debate over the time of the ape-hominid split and a subsequent debate over the relative worth of genetics, as compared to anatomy, for resolving questions of evolutionary history.[16] Since the two approaches gave markedly different interpretations, it was possible that (a) interpretations based on fossils were wrong, (b) interpretations based on genetics were wrong, or (c) both were wrong. As it turned out, the dates advocated by the geneticists were correct, and the initial dates based

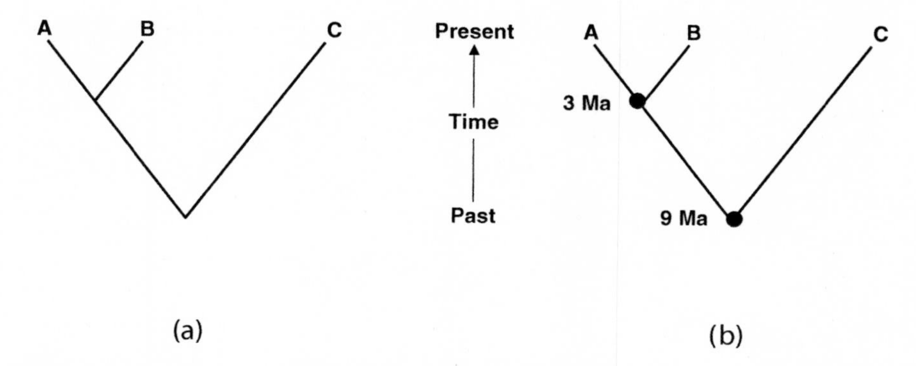

Figure 2.7 The method of molecular dating. (a) Assume that the genetic data show us that species A and B are more closely related to each other than either is to species C, and that the genetic distance separating C is three times that separating A and B. (b) If this is the case, and if we know from fossil data that species C split off from a common ancestor 9 million years ago (Ma), then A and B split from a common ancestor 9 ⅓ = 3 Ma.

on the fossil record were incorrect. This should not be read as a triumph of a molecular genetic approach to an anatomical approach, because in the long run what convinced most anthropologists was not the molecular dating but rather the accumulation of more fossil evidence.

Subsequent research on *Proconsul,* including postcranial remains (the skeleton below the skull), showed that it was actually a rather primitive ape with a mixture of ape-like and monkey-like features. It lacked a tail, as do modern apes, but had arms and legs roughly equal in length, a trait found in monkeys but not in apes, who have longer arms than legs. The entire skeleton shows a mosaic of features and is perfectly consistent with the beginning of the entire hominoid line soon after it split from the Old World monkeys. The close similarity in teeth between *Proconsul* and living African apes is simply an example of primitive traits that were retained in descendants. In addition, further research on *Ramapithecus* showed that it was definitely an ape and not a hominid.

Since Sarich and Wilson's initial work, a large number of studies have used molecular dating based on a variety of data, including DNA sequences. Although the specific estimates often vary depending on what genetic measures are used, the results are very consistent (Figure 2.8). The orangutan is estimated to have split from the African line about 12 to 16 million years ago. Although there is still debate regarding the evolutionary relationships among the African apes, the evidence points to a

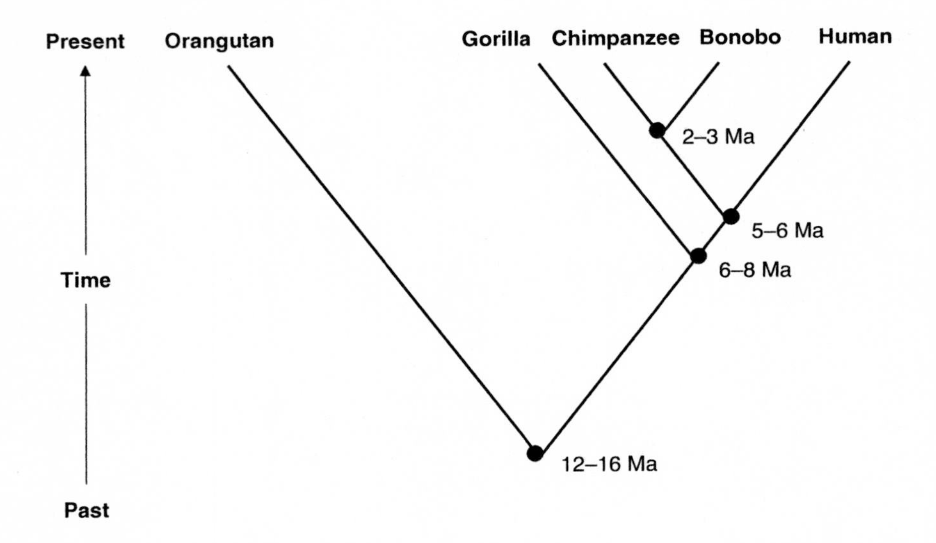

Figure 2.8 The evolutionary history of humans and great apes. Estimated dates are from molecular dating. Genetic estimates that the hominid line split off 5 to 6 Ma is very consistent with the fossil record of the first hominids, now known to have lived almost 6 million years ago.
Source: Horai et al. (1995); Gagneux et al. (1999); Chen and Li (2001).

slightly earlier divergence of the gorilla line at about 6 to 8 million years ago, and then a split between the chimpanzee-bonobo line and the hominid line at about 5 to 6 million years ago.[17] These dates are consistent with what we currently know about the fossil record, particularly for our lineage, where the earliest evidence of bipedalism dates close to 6 million years ago.[18]

What Is the Real Family Tree?

A quick reading of the history of primate classification and evolution might lead one to conclude that genetics provides the correct answer and that anatomical traits do not. Indeed, the debate over genetic and anatomical data for reconstructing evolutionary history is sometimes portrayed in this manner, often explicitly stated in terms of the superiority of genetic data. Such statements are misleading because it is not a matter of genetics versus anatomy, but rather a matter of *which* anatomic traits are used. If we go beyond aspects of overall physical similarity (such as the "naked" appearance of humans relative to the great apes), there is evidence that anatomy also shows the close kinship of African apes and humans. As early as the nineteenth century, both Charles Darwin and Thomas Henry Huxley interpreted anatomical variation as showing greater similarity between humans and African apes than between either and orangutans. The same conclusion has also been reached by other scientists, including a recent analysis of soft tissue anatomy (focusing on the muscular, vascular, and nervous systems).[19]

What actually matters when comparing different traits is the extent to which variation in a trait reflects evolutionary kinship as opposed to the extent to which it reflects unique adaptations. One way of dealing with this is to determine whether a given trait seems to be primitive (of ancient origin) or seems to be derived (of recent origin). Primitive traits don't tell us much about evolutionary relationships because they are shared by descendant species. Humans don't have tails, but neither do any of the apes. This trait (absence of a tail) is shared among all hominoids. Although it is a useful trait in showing how hominoids, *as a group,* are different from other primates, it can't be used to determine whether African apes are more similar to humans or to orangutans, because they all share this trait. Lack of a tail is a primitive trait within the hominoids (although it is a derived trait relative to other primates; the categorization of traits as primitive or derived is relative to the specific set of species being compared). Derived traits provide more information if they are shared by two or more species, because the simplest explanation for shared derived traits is an evolutionary connection between species. For example, the African apes are all closely related because they all share a knuckle walking anatomy, a derived trait. Knuckle walking is most likely shared by these three species through inheritance from a common ancestor rather than each species having evolved this trait independently.

A branch of biological classification known as cladistics is based on the principle that the sharing of derived traits, as opposed to primitive traits, tells us significantly more about the evolutionary relationships among different species. Complex methods are used to determine the pattern of many different derived traits among species. These methods confirm the close evolutionary relationship between humans and African apes.

Genetic and anatomic analyses show African apes and humans to be more closely related, in terms of evolutionary history, than either is to the orangutan. However, it is also obvious that humans are quite different in several adaptations. We are relatively hairless, have large brains, have small canines, and walk upright. These are all examples of derived traits that are *unique* to the human line. In terms of reconstructing evolutionary history, we would not focus on such traits because they don't provide any information about shared traits. That is, they can't tell us whether we are more similar to chimpanzees than to gorillas; they simply tell us that we are different.

Putting these ideas together allows us to reconcile our close relationship with the African apes with the obvious unique features that we have. Although we share close kinship with the African apes, some of our features have changed dramatically since the time of our common ancestor. When we look at humans and great apes in terms of traits such as bipedalism or brain size, we are seeing a demonstration that we have changed more and they have changed less in some features. On the one hand, we are very similar to the African apes, and on the other, we are different. It depends on what features we are looking at.

Hominids or Hominins?

All of the discussion so far can be summarized with two basic points. First, our closest living relatives are the African apes. Humans and African apes are more closely related to each other than either is to the Asian great ape, the orangutan. Indeed, it now appears that humans and some African apes (chimps and bonobos) are more closely related to each other than either is to the third African ape, the gorilla. Second, examination of unique derived traits in humans shows that in some ways we have changed dramatically from our common ancestry with African apes.

In terms of classification, what should we call ourselves? Should we focus on our similarity with the African apes or on our differences? The traditional classification scheme given in Figure 2.1 emphasizes our differences from the other living hominoids. In terms of our bipedal stance, our large brain, our small canines, and other traits, this is a reasonable classification. The problem here is that it doesn't fit the actual pattern of overall genetic and evolutionary relationship, which indicates that humans should be grouped more closely with the African apes. In other words, the traditional classification does a good job of showing differences in adaptation but fails to reflect existing genetic relationships.

So, should we classify ourselves according to our overall genetic and evolutionary relationship to other species, or should we focus on those traits that have changed uniquely in the course of human evolution and make us different? Should classification reflect common evolutiona[illegible] tory or unique adaptations? There continues to be a great deal of debate on this issue. Recent proposals have formed classifications based largely on genetic relationships. One example is shown in Figure 2.9. Here, only two families are recognized within the superfamily of hominoids: the hylobatids (gibbons) and the hominids, where hominid is redefined to include the great apes and humans. The hominid family is then broken down into three subfamilies, corresponding to orangutans, gorillas, and a group known as hominines, which includes chimpanzees, bonobos, and humans. The hominines have two subgroups, known as tribes, one of which includes the chimpanzee and the bonobo, and the other, known as hominins, which consists of humans. The hierarchy shown in Figure 2.9 replicates the tree of genetic relationship shown in Figure 2.6. This classification is useful in showing our evolutionary relationships with other species, but it doesn't show our unique adaptive differences as well as the traditional classification.

This new classification system has the obvious advantage of being congruent with what we know about the actual evolutionary relationships of the hominoids. This scheme, or variations on it, is being proposed by a number of scientists as a way to make our classification system consistent with the evolutionary relationships gleaned from studies of genetics and derived traits. From this perspective, one can make a good zoological argument that we are essentially "naked apes."

Of course, the traditional system has been in place for quite some time, and the fact that conventional labels are used in quite different ways can be more than a little confusing. The major problem lies with the term "hominid." Traditionally, this term has been used to define living humans and other past species on the evolutionary line that split from the African apes; it was defined most often in terms of bipedalism. Under the new system, the term "hominid" continues to include humans (and human ancestors) but also includes chimpanzees, bonobos, gorillas, and orangutans! In other words, in one system, hominid is synonymous with humans, and in the other, hominid means humans *and* great apes. This is understandably confusing to the novice or beginning student!

Although new ideas are often confusing, this is not a sufficient reason to keep things the way they are. If there were general agreement on the new system, then eventually all textbooks, encyclopedias, and lectures would adopt the new use of the term "hominid." It would be a little more

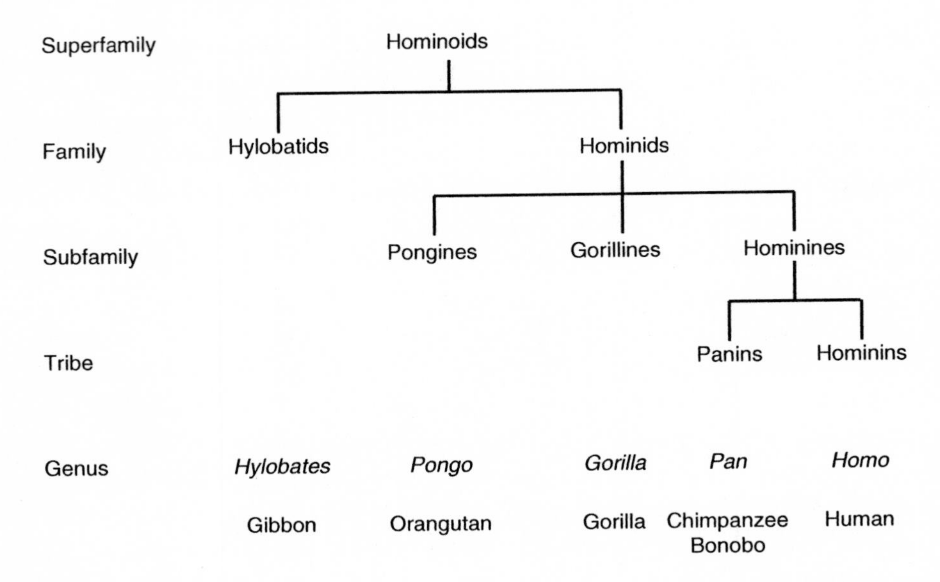

Figure 2.9 What are humans? One of several revised taxonomic schemes for hominoids designed to better reflect the genetic evidence of close kinship between African apes and humans. Compare this to the classification presented at the beginning of the chapter. This classification more accurately portrays evolutionary relationship, whereas the classification in Figure 2.1 emphasizes the unique adaptations of humans.

challenging when reading the older literature, but the confusion could be minimized with appropriate historical background.

So why do many scientists still use the traditional system? Is it simply conservatism and an unwillingness to rewrite one's lecture notes? For one thing, although many are dissatisfied with the traditional classification system, it is not yet clear what the best alternative might be. Some would argue for the system in Figure 2.9, but others would not. Although there is some agreement that the term "pongid" has outlived its usefulness, it is less clear how to fix the problem. Other arguments revolve around the appropriate taxonomic rank for humans—family, subfamily, tribe, or something else. An international conference is scheduled for 2003 to address some of these issues.[20]

There seems to be a growing consensus that whatever the final revision in classification, it should be a scheme that reflects the evolutionary relationships now apparent to us. Although I have no particular argument with this approach, it is only one approach. As noted at the start of this chapter, the "correct" classification system depends on the specific tions one is trying to answer. The system shown in Figure 2.9 and variations on it are perfectly appropriate *if* the purpose of a classification system is to reflect evolutionary relationships. Thus, humans, chimpanzees, and bonobos *should* be placed in a group that emphasizes their kinship *if* the goal of classification is to illustrate this kinship. On the other hand, a case can be made for a classification system that emphasizes *adaptation,* where humans should be considered separate from the African apes, and the family status of "hominid" reflects this.

Which system is preferable? It depends on what you are trying to show. To some extent, this question reflects a variety of philosophical stances as well as scientific insights. Personally, I am sympathetic to organizing classification systems around the principle of evolutionary relationships, but I am not convinced that these relationships alone should determine our taxonomic status. Yes, we are very closely related to the chimpanzee and bonobo, and we could be considered "naked apes," as Desmond Morris called us, or "the third chimpanzee," as Jared Diamond called us.[21] Nevertheless, it is also clear that our species has changed both anatomically and behaviorally. Whether we acknowledge this in our formal classification depends on how much we wish to stress evolutionary relationship and how much we want to stress differential adaptations. Either way, the genetic evidence has clearly shown that we are very similar to our closest living relatives.

THREE

Do You Know Where Your Ancestors Are?

Imagine it is 150,000 years ago. Do you know where your ancestors are?

This question is an obvious rip-off of the old public service announcement, "It is 10 P.M.—do you know where your children are?" but it gets to the heart of what is known as the modern human origins debate. All humans throughout the world today are what we call anatomically modern. In addition to a large brain, modern humans generally have a fairly well-rounded skull with reduced brow ridges and a prominent chin. The first appearance of modern humans in the fossil record occurs in different parts of the Old World (Africa, Asia, and Europe) between about 130,000 and 30,000 years ago. The fossil record shows that earlier humans ("archaic humans") also had a large brain, but with a differently shaped skull that was lower and longer. Archaic humans also lived across the Old World. It is clear in a general sense that some population(s) of archaic humans evolved into modern humans. Which one? There are two basic views on this, one proposing that many (perhaps all) archaic populations were part of our ancestry, and the other proposing that modern humans arose from archaics in only one place, Africa, in the past 150,000 years or so.

I find it is easier to understand this debate by framing it in terms of our ancestry. We definitely have ancestors, and they definitely lived in different places at different times. Where did our ancestors live 150,000 years ago? According to one view, *all* of our ancestors alive at that time lived in Africa. It doesn't matter where your more recent ancestors lived in the past few thousand years, be it Europe, Africa, or elsewhere; if this model is correct, and you could trace your ancestry back over thousands of generations, you would find that each and every ancestral line goes back to Africa no more than 150,000 years ago. Although this view of a single recent African origin of modern humans has attracted many supporters over the past few years, it is by no means universally accepted. Others argue a different interpretation, where *some* of our ancestors 150,000 years ago lived in Africa, but others lived elsewhere in the Old World.

What makes this debate particularly interesting to me are the implications for our understanding of the fossil record. We know that by 150,000 years ago, large-brained archaic humans lived throughout the Old World. These archaic humans had brains roughly equal in size to our own today, but with a differently shaped skull. Today, we have modern humans living across the entire planet. What is the relationship of the archaic humans to modern humans? Again, it boils down to a question of *which* archaic human populations are ancestral to us—just those in Africa, or those from more than one region? If all living humans had only African ancestors 150,000 years ago, then what happened to the closely related archaic humans that were living outside of Africa, such as the enigmatic Neandertals of Europe and the Middle East? Were they a different species? If they left no descendants, then why did they die out? If, on the other hand, the transition from archaic to modern humans took place in more than one continent, then how did these different populations interact over time, and can we determine how much of an ancestral contribution each group made?

The focus here is on the evolutionary history of the human species over the past few hundred thousand years. Although this debate has relied on information from the fossil and archaeological records, genetic data have also come into play. This chapter examines how genetic variation in *living* humans informs us about the history of our species, specifically the question of *who* our ancestors were. We use data from living human populations to provide us with a reflection of past events.

A Quick Summary of Human Evolution

At the time of the initial publication in 1859 of Darwin's *On the Origin of Species*, very little was known about the fossil record of human evolution. Since that time, there have been many discoveries that have filled in the basic story of human evolution. Although the fossil record of human evolution is more complicated than once thought and many questions remain, the general picture is well known. As discussed in the previous chapter, living humans are classified as hominids (or hominins by some), a group that also includes our ancestral kin since the split of the African ape and hominid lines roughly 6 million years ago. The major defining characteristic of hominids is that they are bipeds (upright walkers). To date, the oldest known forms suggested to have been hominids include *Ardipithecus,* dating back to 5.8 million years ago, and *Orrorin,* dating back to 6 million years ago. At present, there is a growing debate over their evolutionary relationships to later forms. At the time of this writing, another species, *Sahelanthropus tchadensis,* has been announced as a possible early hominid, dating to between 6 and 7 million years ago.[1]

Despite this debate, several points of agreement have emerged. Early hominids lived in Africa close to 6 million years ago. They appear to have been bipeds but with small ape-sized brains. In some ways, we can consider these early forms bipedal apes: human ancestors, but not human, at least in terms of traits we usually associate with being human, such as a larger brain and the possession of stone tool technology. The date, location, and anatomy of these fossils is consistent with other evidence on human evolution. We know from both genetic and anatomical studies that the African apes are our closest living relatives, and molecular dating suggests our line split from theirs roughly 6 million years ago, which is consistent with the fossil record. Given that African apes are, by definition, from Africa, it makes sense that the first hominids would also be found in Africa (all hominids older than 2 million years old have been found only in Africa). If we consider the differences between living African apes and humans to have emerged over the past 6 million years, then we would expect that the farther back in time we look, the harder it should be to tell the lines apart. Evolutionary theory predicts exactly what we observe: The first hominids are very ape-like in many features.

For the next 4 million years, hominid evolution took place exclusively in Africa. There were many different forms of early hominid, including a number of species in the genus *Australopithecus,* who walked upright but still had a small brain and a large face. There appears to have been a great deal of variation between 3 and 2 million years ago, with different species developing different sets of adaptations, such as large back teeth among several species. Although the teeth become less ape-like over this time period, there is little change in brain size and no evidence of stone tool technology until about 2.5 million years ago.[2]

Things begin changing rapidly in Africa sometime around 2 million years ago with the emergence of the genus *Homo,* characterized by a larger brain and a stone tool culture. Current evidence suggests several species of hominids with larger brains lived in Africa at this time, of which one, *Homo erectus,* is our ancestor (Figure 3.1a). *Homo erectus* is the first hominid to expand out of Africa, reaching Southeast Asia and Eastern Europe about 1.7 million years ago.[3] Although their brain size was somewhat smaller than ours today, *Homo erectus* had an essentially human skeleton from the neck down, made sophisticated stone tools, and possibly used fire. From this point on, we find evidence of early humans in various places throughout the Old World. Brain size continues to increase over time, particularly after about 700,000 years ago.

Although large-brained like modern humans, the archaic humans were different from us in several features of their anatomy. Compared to modern humans, their skulls were generally lower and longer, with sloping foreheads and large brow ridges and faces (Figure 3.1b). Some archaic humans, such as the Neandertals of Europe and the Middle East, were even more distinct, with large noses and protruding midfacial regions (Figure 3.1c). Anatomically modern humans (Figure 3.1d) begin to appear in the fossil record first in Africa around 130,000 years ago and then later throughout the rest of the Old World. The moderns generally had a more rounded skull, a more nearly vertical forehead, smaller brow ridges, and a noticeable chin.

How Many Species?

At the simplest level, we can describe the evolution of the genus *Homo* over the past 2 million years as a change from early humans *(Homo erectus)* to archaic humans to modern humans. Although accurate, this simple model does not convey the specific nature of these changes over time. Much of the debate over the evolution of the genus *Homo* centers on the number of species that are represented in the fossil record.

Some anthropologists see the fossil record of *Homo* over most of the past 2 million years as representing the evolution of a single species over time (Figure 3.2).[4] The names "*Homo erectus,*" "archaic humans," and "modern humans" are viewed simply as labels along a continuum of change and are not meant to correspond exactly to separate biological species. An analogy is the growth and development of a human being. We start out as infants and then change into children, preteens, teenagers, young adults, middle-aged adults, and elderly. We recognize distinct

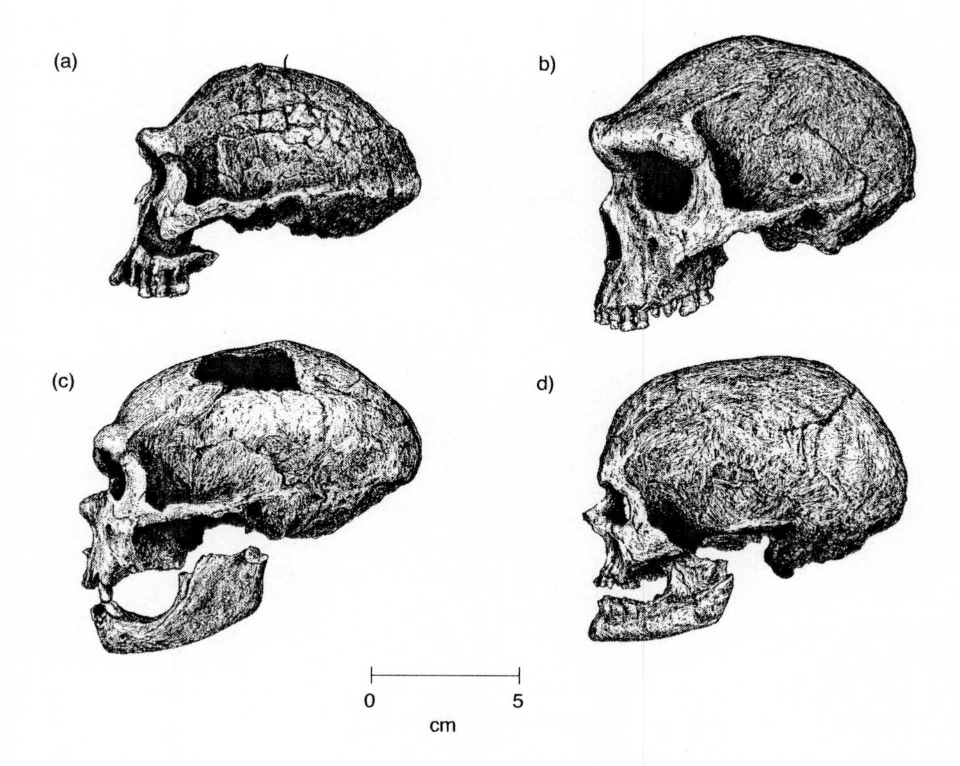

Figure 3.1 Some fossil hominids in the genus *Homo.* (a) The skull of *Homo erectus* specimen KNM-ER 3733 from Kenya, dating to 1.8 million years (b) The skull of an archaic human, the Broken Hill 1 specimen from Kabwe, Zambia, dating to roughly 300,000 years ago (c) The skull of a Neandertal, the La Chapelle-aux-Saints specimen from France, dating to roughly 40,000 years (d) The skull of an early modern human, the Cro-Magnon 1 specimen from France, dating to roughly 23,000 to 27,000 years. Reprinted, by permission, from C. S. Larsen, R. M. Matter, and D. L. Gebo, *Human Origins: The Fossil Record,* third edition (Prospect Heights, Ill.: Waveland Press, 1998).

stages in our lives, but the boundary between them is often arbitrary. Regardless of what labels we use, each of us constitutes only a single person at any stage of our lives. Likewise, terms such as "early humans," "archaic humans," and "modern humans" represent different stages in the ongoing evolution of a single species.

The simplicity of this model is appealing in many ways. However, its simplicity does not mean it is correct. A growing number of anthropologists view the fossil record of *Homo* rather differently, claiming evidence of two (or more) species of humans at any particular time.[5] An extreme

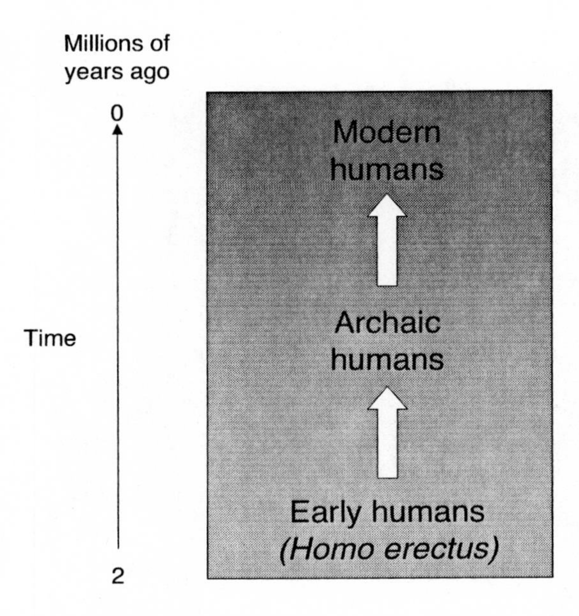

Figure 3.2 A view of human evolution over the past 2 million years proposing that all human fossils belonged to a single evolving species. The boundaries between early humans, archaic humans, and modern humans are considered somewhat arbitrary. Compare this view with the model shown in Figure 3.3.

version of this view is shown in Figure 3.3. Here, what others call *Homo erectus* (or early humans) is divided into two species; an initial species in Africa, known as *Homo ergaster,* gives rise to at least two species—*Homo erectus* and *Homo antecessor.* Here, the label *Homo erectus* is given to an Asian branch of early humans that ultimately became extinct somewhere between 30,000 and 200,000 years ago. *Homo antecessor* evolves into *Homo heidelbergensis,* which in turn splits into two species: *Homo neanderthalensis* (the Neandertals), who become extinct, and *Homo sapiens,* who are modern humans. According to this view, the fossils that others call "archaic humans" actually represent a number of different species. Such models generally recognize three human species in existence 100,000 years ago: surviving populations of *Homo erectus* in parts of Asia, Neandertals in Europe and the Middle East, and our own ancestors, the first *Homo sapiens,* arising in Africa.

Species identification is a tricky business. For one thing, there are different definitions and conceptual models relating to exactly what is meant by "species." Most evolutionary biologists use what is known as the bio-

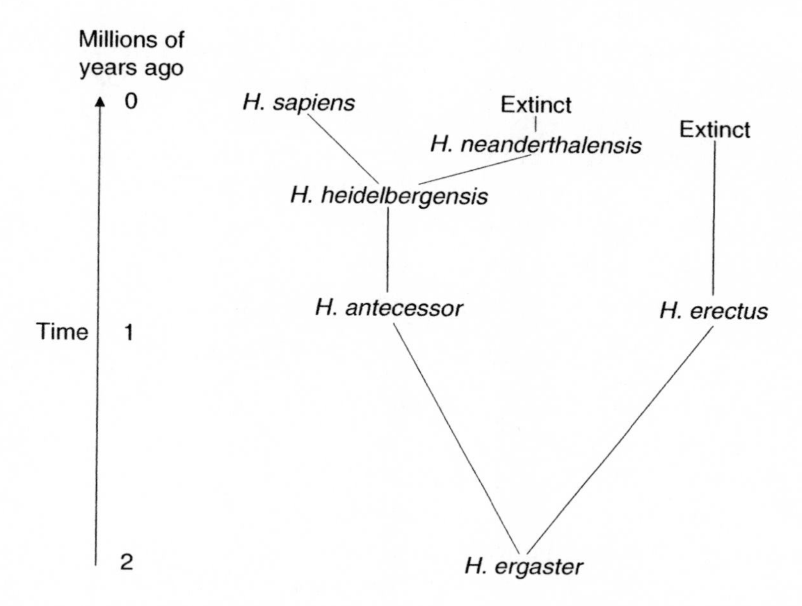

Figure 3.3 A view of human evolution over the past 2 million years proposing that there were a number of distinct species that lived at different points in the past. Compare this view with the model shown in Figure 3.2.

logical species concept, the idea that two populations belong to the same species if they naturally interbreed and produce fertile offspring. Among living humans, fertile offspring result no matter what part of the world the parents are from; thus, all living humans belong to a single species. Elephants and spiders are obviously different species because they do not interbreed. Horses and donkeys do interbreed, but their offspring (mules) are sterile, and thus they are classified as separate (albeit closely related) species.

Although useful when dealing with living organisms, the biological species concept is difficult to apply to fossil remains, where we have no direct evidence on interbreeding or fertility and must instead make inferences based on overall physical appearance. In some cases, this is not difficult. For example, the fossil record in East Africa shows the coexistence of *Homo erectus* with a species of *Australopithecus* that is quite different, having a small brain and very large back teeth. The situation is trickier when dealing with archaic and modern humans. In some ways, such as brain size, these forms are very similar, but in other ways, such as cranial shape, they tend on average to be different. The basic question is *how* different—

enough to classify them into different species? Some anthropologists emphasize the similarities, whereas others focus on the differences.[6] There is no perfect solution to this problem, and it is compounded by personal views regarding the nature of variation and speciation. Some, by virtue of specific training and evolutionary philosophy, tend to be "splitters," recognizing many species, whereas others tend to be "lumpers," placing observed variation within a single species.

The Origin of Modern Humans

Differing interpretations of the number of past human species spring from the debate over the origin of modern humans. Many different approaches have been taken to frame this debate. Here, I examine two basic models: African replacement and multiregional evolution.

The major difference between the African replacement model and the multiregional evolution model lies in the question of *where* our ancestors lived some 150,000 years ago, the period just preceding the earliest fossil evidence of anatomically modern humans. According to the African replacement model, *all* of our ancestors came from Africa. Humans living outside of Africa at this time (such as the Neandertals) were not our direct ancestors but were cousins of a side branch of our family tree who eventually became extinct. In contrast, the multiregional evolution model holds that while *some* of our ancestors lived in Africa, others lived outside of Africa, so that our ancestry today includes some genetic contributions from populations in more than one continent.

Figure 3.4 illustrates the African replacement model (also known, in various papers, as the "recent African origin model," the "out of Africa model," the "Garden of Eden model," and the "Eve model," among others). This diagram illustrates evolutionary connections across time and space. From the fossil record, we know that archaic humans lived in parts of Europe, Africa, and Asia. We also know that modern humans lived in all of these regions for at least the past 30,000 years or so. How do these different groups relate to one another? According to the model of complete replacement, the transition from archaic to modern took place in Africa *and only in Africa* roughly 150,000 to 200,000 years ago. A new species, *Homo sapiens,* split off from an earlier archaic human species. By 100,000 years ago, populations of this new species began to disperse out

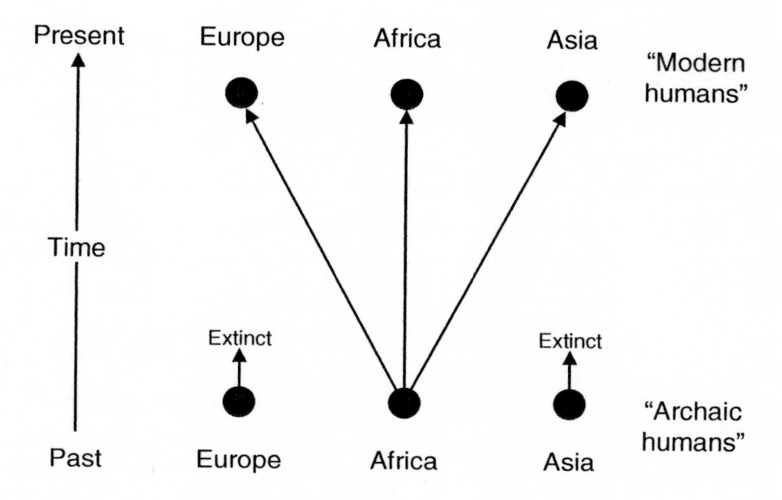

Figure 3.4 The African replacement model of modern human origins. According to this model, the transition from archaic to modern human took place first in Africa 150,000 to 200,000 years ago. Archaic humans outside of Africa became extinct and did not contribute to our ancestry. All of our ancestors 150,000 years ago lived in Africa. Compare this view with the model shown in Figure 3.5.

of Africa, first into the Middle East and then later into Australia, Asia, and Europe. Meanwhile, there were still populations of "archaics" living outside of Africa. According to the replacement model, the non-African archaic populations were eventually replaced by newly arriving modern populations from Africa, and consequently, all living humans trace *all* of their ancestry back 150,000 years or so to the initial appearance of *Homo sapiens* in Africa.

If this did indeed happen, then what happened when the new modern humans met the archaic humans? Was this meeting peaceful? Violent? Did they ignore each other? Various replacement models have been proposed, ranging from those suggesting that moderns were better adapted for speech to those arguing for higher fertility among the moderns. There does not appear to be any significant evidence of violence, unlike the visions of invading Cro-Magnons conjured up in stories such as William Golding's novel *The Inheritors.* We are still debating whether replacement occurred, and it may not be very productive to argue much about how it happened before we know whether it actually occurred. The African replacement model posits that there was no genetic contact between the archaics and moderns. Perhaps they did not mate with one another. Perhaps they did but were so genetically different that they did not produce fertile offspring, much like the case of horses and donkeys.

The multiregional evolution model takes a different view of the genetic relationships between archaics and moderns. Here, different populations are all part of the same species, and human populations in different geographic regions are interconnected genetically by migration. A genetic change occurring in one population is ultimately shared by the movement of genes into other populations, a process known as *gene flow,* which happens when an individual mates with someone in another population.

Gene flow connects the gene pools of two populations, and it can happen in two ways. First, there may be direct gene flow between the two populations, where an individual moves into a different population and has offspring with a mate in that new location. Second, two populations can be connected via a network of intervening populations, where genes flow from one population to the next over many generations. For example, imagine you have three populations (1, 2, 3) arranged geographically in a straight line: 1–2–3. If someone from population 1 mates with a resident of population 2 and has offspring, then gene flow has occurred from population 1 into population 2. If, in the next generation, someone with those genes in population 2 mates with a resident of population 3 and has offspring, the genes originally in population 1 have now made it into population 3. Over time, gene flow can spread genes across a wide range. In the real world, the process is likely to be more complicated, as gene flow could also take place in the reverse direction as well.

Evolutionarily, gene flow does two things. First, it acts to introduce new genes into a population. Second, it acts to reduce genetic differences between populations; the more two populations mix, the more similar they will be genetically, much the same way as mixing two cans of paint will make the colors in each can similar. A key point of multiregional evolution, with its emphasis on gene flow, is that our genetic diversity in the world today has resulted from a mixture of genes from different parts of the world over the past several hundred thousand years or more. According to this model, some of our ancestors 150,000 years ago did live in Africa, but others lived elsewhere. Most advocates of multiregional evolution suggest that the same process marks the entire time span of the genus *Homo,* going back close to 2 million years ago. Following the dispersal of

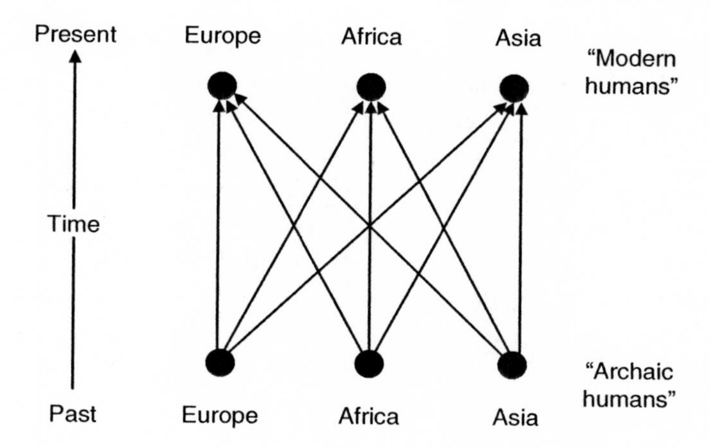

Figure 3.5 The multiregional evolution model of modern human origins. According to this model, some of our ancestors 150,000 years ago lived in Africa, but others lived outside of Africa. Archaic human populations outside of Africa contributed to our ancestry. Compare this view with the model shown in Figure 3.4.

some early humans *(Homo erectus)* from Africa, human populations in different parts of the Old World have remained connected via gene flow in a single species that evolved over time (Figure 3.5). Some changes occurring in one part of the world were ultimately shared elsewhere. This does not mean that all populations were identical. Some evolutionary forces act to increase geographic differences, but gene flow acts to counter these sufficiently to prevent a new species from splitting off.

There are different variants of the multiregional evolution model. One view, which I call the "primary African origin model," agrees with replacement advocates that the major genetic and anatomic changes from archaics to moderns *did* take place first in Africa but were then shared with other archaic groups outside of Africa through the action of gene flow.[7] Whether this would have happened through the direct movement of people out of Africa across long distances or through the step-by-step process of gene flow from one population to the next, and then to the next, and so on, is not clear. The basic idea, however, is that the genes for modernity spread out of Africa into the rest of the Old World. Mixture, rather than replacement, characterized the transition out of Africa.

Another view on multiregional evolution, which I call the "regional coalescence model," suggests that the genetic and anatomical changes in the transition from archaics to moderns did not all happen at a single place or time. Instead, some changes started in one part of the world, such as Africa, and others took place elsewhere, such as parts of Europe or Asia. Each of these changes was ultimately shared with humans in other parts of the world through gene flow, and the transition from archaic to modern resulted from the coalescence of all these changes. As argued below, I don't think that this view of multiregional evolution is the correct interpretation. However, I am not convinced that complete replacement took place either.

The Fossil Record

Although this book deals with the use of genetic data for reconstructing human history, it is worthwhile to examine briefly the pros and cons of the different origin models in terms of the fossil record. Although genetic information has many advantages for studying population history, it cannot be considered in isolation from the fossil record, which is, after all, direct evidence of past events.

The first thing we can examine is the distribution of human fossils over time and space to determine when and where modern humans first appear. This is not as easy as it sounds. One problem is the incompleteness of the fossil record. We have fossil data on only a tiny fraction of all individuals who ever lived because of many factors influencing preservation and recovery. Thus, we must always be alert to new discoveries that help fill in the gaps across both space and time. Another problem is reaching consensus on exactly what constitutes "archaic" and "modern." Some anthropologists argue that this distinction is rather arbitrary,[8] much like the difference between someone who is middle-aged and someone who is elderly, and that our efforts to force specimens into one category or the other might bias our results. This is certainly the case for fossil specimens that some call "early moderns," in that they have some features usually associated with modernity, such as a vertical forehead or a noticeable chin, but also have some archaic features, such as large brow ridges. There is understandable disagreement about such specimens. There is also debate over the dating of some key fossils.

Nevertheless, many anthropologists agree that a general pattern can be discerned by considering all specimens across both time and space. There

appears to be a growing consensus that anatomically modern
sils appear first in Africa, with the oldest such specimens l
130,000 years old. Although the first African moderns are not
living humans, these specimens have often been interpreted as having enough modern traits to classify them as such. Somewhat younger, although still quite ancient, fossils of moderns have been found in the Middle East that date to about 90,000 years ago. Our evidence for East Asia is still fragmentary, but it appears that Australia was first settled by modern humans as early as 60,000 years ago. Europe is an interesting case, in that fossils identified as Neandertals and fossils of moderns have been found that date to the same time. Modern humans appear in Europe about 40,000 to 30,000 years ago, with some evidence pointing to the younger date in Western Europe, the part of the continent most geographically distant from Africa, the presumed source of the modern populations.[9]

If we accept these dates (and assignments of fossils), an interesting pattern begins to emerge. Modern humans appear first in Africa and then later in other parts of the world. Moderns outside of Africa are found first in the Middle East, which is geographically closest to the African continent. More geographically distant places, such as Australia and Europe, are populated by moderns later in time. This is exactly the pattern we would expect if the African replacement model were correct. This geographic pattern of dates is also consistent with the time it would take for early hunting and gathering peoples to move outward from Africa when movement was limited to walking.

Doesn't this *prove* that the African replacement model is correct? After all, the observed data fit the predictions of the model. Not necessarily. All this shows is that the observed data are *compatible* with the African replacement model. This compatibility will prove the accuracy of the replacement model only if the data are *incompatible* with the multiregional evolution model (and with any alternatives). Compatibility does not necessarily equate with proof.

Although the dates for the appearance of modern humans are compatible with African replacement, they are *also* compatible with the primary African origin version of the multiregional model, which postulates an initial change in Africa followed by gene flow outside of Africa. Any movement of genes, either through the steady flow from population to population through interbreeding or by the physical movement of groups of people, is going to take time. Moreover, the amount of time needed for gene flow would be related to the geographic distance from Africa—first to the Middle East and later to other parts of the Old World. Both the African replacement model and some versions of multiregional evolution predict the same pattern, and therefore we cannot distinguish between them based on the dates.

A key prediction of multiregional evolution (all versions) is that the fossils will show what is known as *regional continuity,* the persistence of some traits over time in the same region. As an example, consider the case of the horizontal-oval mandibular foramen (try saying that three times fast!). There is an opening (foramen) on the inside of the vertical part of your lower jaw (mandible) through which a nerve passes. There are different forms of this opening. In most humans, there is a long groove connected to the bottom rim of the opening. In others, the groove is covered and the opening is oval-shaped with the long axis of the oval horizontal to the jaw. As far as we know, this is a neutral trait, so that whatever form you have has no effect on your survival. The horizontal-oval form is found among European Neandertals, an archaic human population, where over half the jaws have it, and is rarely seen in archaic populations outside of this group. This same trait is also found in European post-Neandertal modern humans, although at a reduced frequency. The continuity of this trait over time in Europe has been argued as evidence that the European archaics made *some* genetic contribution to later modern Europeans.[10] This is expected under multiregional evolution, but it is harder to explain under a model of African replacement, where moderns came from a population in Africa that, as far as we know, lacks this trait, since we assume that independent evolution of such traits is rare. How then did the trait reappear in Europe?

A number of anatomical traits have been proposed as evidence of regional continuity, with particular emphasis on measurements of cranial and facial shape in East Asia and the region around Southeast Asia and Australia. Much of this evidence has been gathered through the efforts of Milford Wolpoff of the University of Michigan and his students. Although a number of anthropologists have accepted evidence for regional continuity (and thus a multiregional interpretation of modern human origins), there is still considerable debate over the data (often complicated by the fragmentary nature of the fossil record) and alternative explanations.[11] For one thing, not all traits are likely to be equally informative about population history, particularly those that are shaped by natural selection and adaptation to a local environment. Among living humans, a good example would be the similar dark skin color of populations in Central Africa and in highland New Guinea. The fact that these groups are similar in skin color does not imply a common recent history, because skin color is strongly affected by adaptation to local environments, such that populations living in or near the equator have adapted to high levels of ultraviolet radiation by having darker skin. Traits that are strongly affected by selection and adaptation to different environments can confuse the issue. An example from the fossil record are populations that share a pattern of relatively stocky bodies and proportionately shorter limbs, a feature that is likely to arise in populations adapting to cold climates, since body and limb shape appears to influence heat retention.

However, there is evidence for some continuity in some parts of the world outside of Africa. If we combine this observation with the known distribution of archaics and moderns over time and space, the most logical model (to me) would embrace an African origin of modern humans *combined* with gene flow outside of Africa. Recent analyses have compared a number of archaic and modern fossils and found strong evidence that anatomically modern humans generally show a pattern consistent with the view that all modern humans have some ancestors in Africa and others from outside Africa.[12]

The debate over the fossil record is much more involved than presented in the short review here. The purpose of this chapter, however, is to consider the impact that another set of information brings to the debate—the reconstruction of the history of our species based on genetic data from living populations. What does our current genetic diversity tell us about the past?

"Mitochondrial Eve"

For many decades, anthropologists and geneticists have looked at the global pattern of human genetic variation in an attempt to tell something about the past events that shaped this diversity. In general, these early attempts did not have a major impact on those studying fossils. Anthropologists who studied the living and those who studied the dead did not interact very often. This situation changed dramatically in 1987 with the publication in the journal *Nature* of a paper titled "Mitochondrial DNA and Human Evolution," by Rebecca Cann, Mark Stoneking, and Allan Wilson.[13] This journal article presented the results of analyses performed on samples of human mitochondrial DNA from around the world. As described in the previous chapter, mitochondrial DNA (mtDNA) is inherited solely from one's mother, making it very useful in analyses of population history because there is no mixing of genetic material as in nuclear DNA.

If two people have very similar mtDNA, then they share a fairly recent common female ancestor. People who are more distantly related will have significantly different mtDNA sequences. The reason for these differences is mutation. Because mitochondrial DNA is passed through the mother's line intact, without the shuffling we see in nuclear DNA, the only way two people can have different mitochondrial DNA is through mutation. Over time, mutations accumulate, leading to a greater difference in the mtDNA sequences. In other words, given two people who share a common female ancestor, the greater the elapsed time from that ancestor, the greater the probability that mutations have occurred, and the more dissimilar the two people's mitochondrial DNA will be. This means that we can compare mitochondrial DNA among people and reconstruct patterns of common ancestry and, given an estimate of how quickly mitochondrial DNA mutates, we can further estimate *when* people shared a common female ancestor.

In such an analysis, we start by finding links between the people who are most similar to each other and then add more and more people until all have been linked in the most parsimonious pattern of relationships. In this case, parsimony means looking for the minimal number of mutations needed to join any two people to a common female ancestor. A basic principle of the theoretical models underlying these analyses is that any two individuals have common female ancestors in the past, and the goal is to identify the most recent common ancestor.

In order to understand what this means, look at the imaginary genealogy shown in Figure 3.6. This is a partial family tree for three people—Adam, Alice, and Amy. In this and other kinship diagrams, the triangles correspond to males and the circles to females. We see that Adam and Alice are first cousins because they share a set of grandparents—Clem and Chloe. The mother of Adam (Betty) is the sister of Alice's mother (Barbara). We also see that Amy is a second cousin to both Adam and Alice,

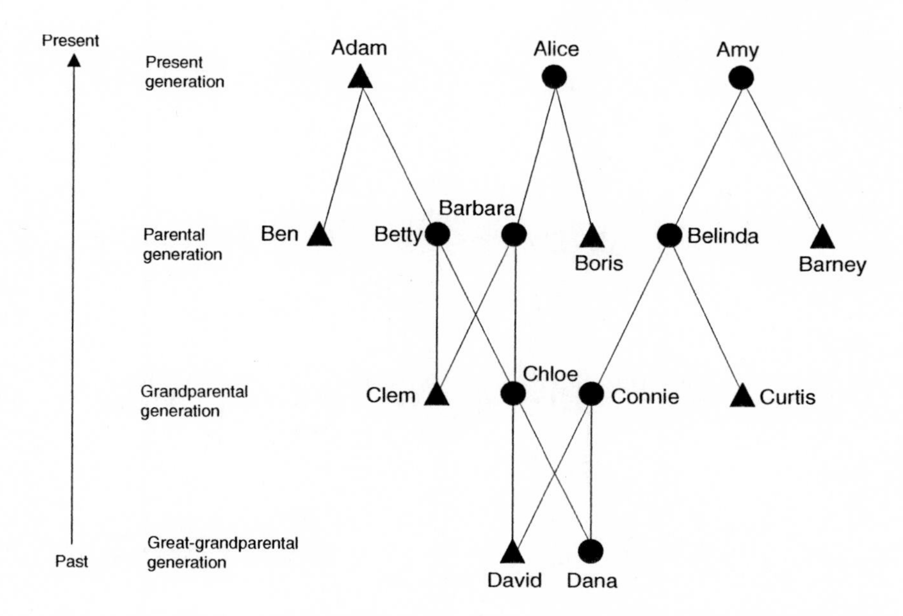

Figure 3.6 A partial genealogy showing the common ancestors of three individuals in the present—Adam, Alice, and Amy. Circles represent females and triangles represent males. Adam and Alice are first cousins because they share a set of grandparents (Clem and Chloe). Amy is their second cousin because they all share a set of great-grandparents, (David and Dana).

since they all share a pair of great-grandparents—David and Dana. As first cousins, Adam and Alice are more closely related to each other than either is to Amy, their second cousin.

The relationships between Adam, Alice, and Amy are clearer when considering only the female line, as would be the case when looking at their mitochondrial DNA. Figure 3.7 shows only the female line of inheritance. The most recent common female ancestor that first cousins Adam and Alice have is their maternal grandmother, Chloe. In turn, Chloe's maternal ancestor (her mother) is Dana, who is also a great-grandmother of Amy. From this perspective, the most recent common female ancestor of Adam, Alice, and Amy is Dana. If we added a fourth person to this analysis and went back far enough, we could locate the most recent common female ancestor of all four. We could then do this for a fifth person, a sixth person, and so on.

We can't reconstruct such a tree in much detail using genealogical information because few of us have complete records going back more

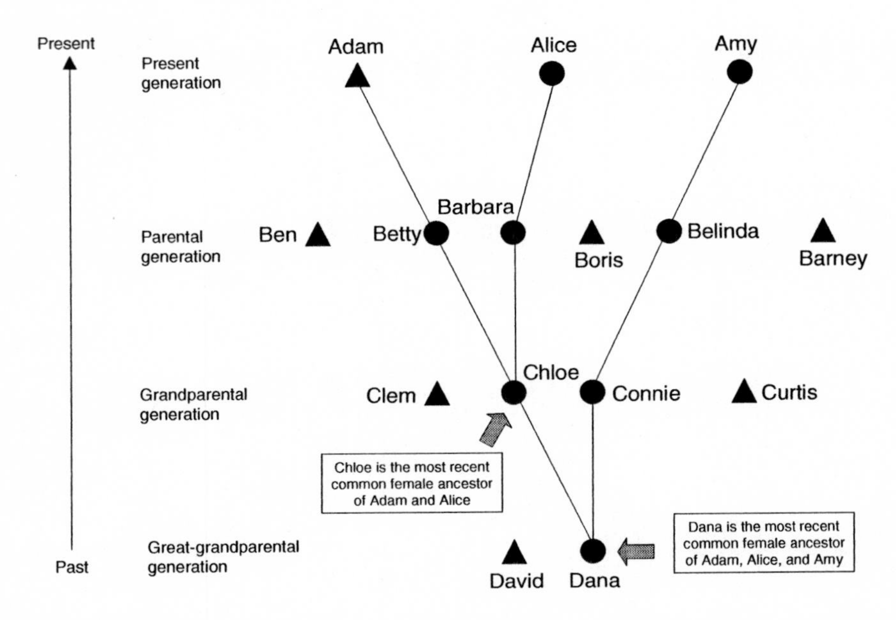

Figure 3.7 The same genealogy shown in Figure 3.6 but showing only the female ancestors. The most recent common female ancestor of Adam, Alice, and Amy is Dana. Analyses of mitochondrial DNA, which is inherited only through the female line, can be used to identify this common female ancestor.

than a few generations. However, we can compare two people's mitochondrial DNA and determine how much they have in common. Then, using an estimate of the mutation rate for mitochondrial DNA, we can estimate how long ago these two people shared their most recent common female ancestor. We could then add a third person, a fourth, and so on, seeking to identify the most recent common female ancestor shared by all of the subjects.

This is essentially what Rebecca Cann and her colleagues did. They sought to identify the common female ancestor of the human species by examining many sequences from people around the world. They looked at the mitochondrial DNA from 147 people representing five different ancestral backgrounds: sub-Saharan Africans; Caucasians (including Europe, North Africa, and the Middle East); East Asians; aboriginal Australians; and aboriginal New Guineans. Some of these people had identical mitochondrial DNA, reducing the number to 133 different mtDNA sequences. The purpose of this analysis was to determine how similar any given mtDNA

sequence was to any other, to use these results to build a "tree" rep
ing the degree of relatedness among the DNA sequences, and to identify the most recent common ancestor. At some point in our past, *all* living humans shared a common female ancestor. When did this woman live? Cann and her colleagues concluded that this ancestral female lived about 200,000 years ago. All living people can trace their ancestry of mitochondrial DNA (although not necessarily other genes) back to this person.

Furthermore, their analysis revealed an interesting pattern of relationship among the mtDNA sequences, which were arrayed in two clusters. The first cluster consisted only of sequences from subjects with African ancestry, while the second cluster consisted of sequences from those in all five ancestral groups, both African and non-African. Because *both* clusters contained sequences of African origin, they inferred that the most parsimonious explanation was that the common ancestor was also African.

The notion of a single common female ancestor elicited biblical allusions, most notably Eve in the creation story in Genesis, which led to the moniker "African Eve" or "mitochondrial Eve." A news article that appeared in the same issue of *Nature* in which Cann's research was published was titled "Out of the Garden of Eden,"[14] leading some to describe the African replacement thesis as the "Garden of Eden" model. This emphasis was a bit misleading because it conjured up an image that all living humans descended from a single breeding pair, Eve and her mate. In reality, the mtDNA analysis identified the time and location of the common *mitochondrial* ancestor, who was only one person out of many who likely lived at that time. Most estimates, based on the relationship between genetic variation and population size, suggest that at least several thousand ancestors may have lived at that time. Of course, mitochondrial DNA allows us to identify only one of them because it is inherited only through the mother's line. Each of us has only one mitochondrial ancestor in any past generation.

The major impact of the analysis conducted by Cann and her colleagues was the implication for the origin of modern human beings. They reasoned that if our common ancestor lived in Africa some 200,000 years ago, then multiregional evolution could not have occurred, because it predicted a much older common ancestor. The observed data were felt to be compatible with the African replacement model and incompatible with the multiregional evolution model. Initial debate over their research focused on a number of technical issues and questions regarding sampling

..s. However, a number of other studies have since appeared that ..m their basic findings.[15] This does not mean that their *interpreta-.ons* went unchallenged.

Certainly, mitochondrial Eve is compatible with African replacement, but is it really incompatible with multiregional evolution? Let's start by considering the location of this ancestor—Africa. We know that there must have been a common mitochondrial ancestor living somewhere at some point in the past. If this ancestor had lived in Asia, then that would obviously weaken the case for an African origin. The reverse is not necessarily true; although an African location of the common ancestor is compatible with replacement, it is also compatible with multiregional evolution. As noted by geneticist Alan Templeton, "Eve" had to live *somewhere*.[16] Under multiregional evolution, this ancestor could have lived in Africa, or in Europe, or in Asia. Knowing the location does not resolve the debate.

The same problem applies to the date. "Mitochondrial Eve" had to live at some point in time. Since she is only one ancestor out of probable thousands, we cannot easily say anything more than living humans have some African ancestry dating to this time. It does not tell us anything about other ancestors. Where did they come from? For that matter, where did Eve's mitochondrial ancestors come from? We can only peer back in time so far using such methods, and this leaves a situation where evidence may not be conclusive. Does the date and location of "Eve" necessarily rule out a multiregional interpretation? Alan Templeton analyzed the geographic distribution of human mitochondrial DNA sequences and concluded that the same results could also be expected under multiregional evolution and that these data did not support the idea of complete replacement.[17] His analytic method is very complex, but it essentially involves looking at the geographic distribution of different mitochondrial DNA types and comparing it to the different expectations he has found for replacement models and gene flow models. His results suggest that our common mitochondrial DNA could have existed in an African ancestor and then spread throughout the Old World by gene flow, mixing with other populations outside of Africa without replacing them.

Again, we are faced with the distinction between compatibility and proof. Had "Eve" been found to have lived outside of Africa, or at a much earlier time, then that would have been strong evidence for rejecting the African replacement model. However, an African origin and a 200,000-

year date are not incompatible with the multiregional evolution model. In fact, such a case is compatible with *both* models.

Gene Trees and Human Ancestry

When geneticists perform an analysis on a single trait, such as a blood type or skin color or a mitochondrial DNA sequence, they are reconstructing the history of that particular trait, which might not be the same as the history of the population that had that trait. For example, recall my discussion earlier in this chapter about skin color in Central Africa and highland New Guinea. Humans in both places have the same trait—dark skin color—but this does not mean that these populations shared a recent common ancestor. Instead, both populations have the same trait because they both adapted in the same way to an equatorial environment. What if mitochondrial DNA was also subject to natural selection? If, for example, certain mitochondrial DNA variants altered one's probability of surviving and reproducing, then our analysis would be showing us the history of these adaptations and not the history of ancestral kinship or gene flow.

Many geneticists argue that this is *not* a problem for mitochondrial DNA. However, this does not mean that we will get a precise reconstruction of population history by relying solely on what mitochondrial DNA can tell us. Different genes or DNA sequences might give somewhat different answers because of chance. Think of this problem in terms of basic probability and sampling. Suppose you have been hired by a manufacturer of athletic footwear to estimate the average shoe size of all male students on my campus. Would you simply come to campus and measure one person? Of course not. Although you might by chance happen to pick someone who had the average shoe size, you might also pick someone whose feet were considerably larger or smaller than the average. You would want to measure a large enough number of people to ensure that you have a good estimate of average foot size. (Statisticians have developed methods to determine exactly how many you would need for a given degree of accuracy.) The same thing happens when pollsters ask the questions you have seen on various news shows. They can't just ask a small handful of people; they need to ask many more to get a reasonable level of accuracy (as expressed by those "margin of error" statistics you have probably seen). In an analogous fashion, geneticists want to sample as many traits as possible to get an accurate estimate of the population history they are trying to reconstruct.

In order to get a good picture of the evolutionary history of our species, scientists have looked at estimates of the location and date of a recent common ancestor using several traits. One useful source of information is the Y chromosome. There are two types of sex chromosomes in humans, X and Y. Females have two X chromosomes (XX), and males have one of each (XY). A child's gender depends on whether the sperm from the father contains the father's X chromosome (which came from the father's mother) or the father's Y chromosome (which came from the father's father). By studying Y chromosomes, geneticists can reconstruct *paternal* ancestry. As with mitochondrial DNA, the Y chromosome is passed along from generation to generation without recombination. (Technically, there *is* a small part of the Y chromosome that does recombine, but most does not, and all discussions in this book refer to the nonrecombining part of the Y chromosome.) In general, the Y chromosome studies also tend to point to a recent common African ancestor, although with an interesting twist. Michael Hammer, of the University of Arizona, and his colleagues have found evidence for movement of the most common Y-chromosome variants out of Africa about 150,000 years ago but also found evidence for movement back *into* Africa from Asia during this time.[18]

Nuclear DNA sequences have also been analyzed by focusing on small sections of the chromosomes that are not subject to the recombination of chromosomes during sex-cell production. The results are mixed. Some of these "gene trees" have roots in Africa, and some do not. Some date back much earlier than "Eve," giving dates of 800,000 years, 1.7 million years, and even earlier.[19] In addition, some genes show evidence of ancient Asian ancestry, going back past 200,000 years, which is difficult to reconcile with a recent complete replacement out of Africa.

Anthropologists and geneticists have been trying to put all these estimates together to get a more accurate view of the history of our species. What has been missing so far is a way to combine results from different gene trees. Most recently, Alan Templeton developed such a method and used it to analyze gene trees for ten different traits (including mitochondrial DNA, Y-chromosome DNA, and nuclear DNA). He found that taken together with fossil evidence, the picture obtained from gene trees is one of multiple dispersals out of Africa. The first such dispersal took place about 1.7 million years ago with the origin of *Homo erectus.* Genetic data suggest a second dispersal of genes out of Africa between 400,000 and 800,000 years ago, and a third dispersal about 150,000 years ago.

Templeton's analysis is significant because it demonstrates that there has been recurrent gene flow among human populations over the past 2 million years. Although his results do suggest that Africa was often the source of new genetic variations, they also demonstrate that replacement was unlikely.[20] The origin of modern humans appears to be out of Africa, but not exclusively so.

Patterns of Human Genetic Diversity

Cann et al.'s support of the African replacement model did not rest solely on the mitochondrial DNA gene tree they were able to construct; they also made an interesting finding about diversity in different geographic regions. There is more diversity of mitochondrial DNA in people with recent African ancestry than in people with recent ancestry elsewhere in the world.

Geneticists measure diversity in a number of ways, including the number and frequency of different genes and DNA sequences. Cann et al. used comparisons of DNA sequences between all pairs of individuals. For example, suppose we are comparing the DNA sequences of four people (A, B, C, and D). We would first compare the sequences of persons A and B to see how many DNA bases differed. We would then compare A with C, A with D, B with C, B with D, and C with D. For each comparison, we would count the number of base differences and then average them over all pairs of comparisons to get an overall measure of diversity. In the Cann et al. study, they found an average of 0.32 percent difference between individuals.[21]

They then looked at comparisons within geographic regions by comparing pairs only with others that had the same ancestral background (comparing Europeans to Europeans, Africans to Africans, and so forth). They found the highest amount of mtDNA diversity in Africans. This finding has been replicated in a number of other studies of mitochondrial DNA, as well as studies of repeated nuclear DNA sequences and studies of physical characteristics, such as cranial measures and skin color.[22] Given the same finding across a wide number of traits, it appears conclusive that the highest levels of genetic diversity tend to be found in sub-Saharan African populations. Why?

Cann and her colleagues interpreted these regional differences in mitochondrial diversity as support for an African replacement model. They inferred that regional differences in diversity are a reflection of a recent

African origin of our species followed by dispersal (and replacement) out of Africa. How can this be? The basic assumption used here is that genetic diversity reflects the age of a population. Mutations introduce new genetic forms and thus increase diversity. As discussed in the previous chapter, mutations accumulate over time. The longer a population has been around, the more mutations have accumulated, and the more genetically diverse a population will be.

Let's tie this in with the African replacement model. In simplest form of the model, modern humans arose in Africa between 150,000 and 200,000 years ago. It is not until about 100,000 years ago that any modern humans moved out of Africa. Therefore, African populations have been around longer, have had more time to accumulate mutations, and show the greatest genetic diversity. Populations outside of Africa were probably first colonized by a small number of individuals (which acts to lower diversity because a small group of people is statistically unlikely to have as much diversity as a larger group). The non-African populations have not been around long enough to accumulate as many mutations and therefore do not show as much genetic diversity. Everything seems to fit together.

One problem with this inference is that it assumes that the African and non-African populations separated and had little subsequent gene flow between them. However, the assumption that Africans, Asians, and Europeans represent separate and independent evolutionary lines extending back in time to a common ancestor does not hold. We observe gene flow between peoples in different parts of the world today, and there is ample evidence that this occurred over long periods of time in the past. When populations mix genes, levels of diversity are affected; the greater the gene flow, the higher the diversity, because gene flow introduces new diversity into a population. Genetic diversity is affected by more than just the accumulation of mutations.

Population Size, Genetic Drift, and Human Evolution

There is also another problem with the inference that genetic diversity reflects a population's age. Another major influence on genetic diversity is population size; smaller populations show less diversity than larger populations because of an evolutionary mechanism known as *genetic drift*.

Genetic drift deals with probability. Imagine, for example, that you have four coins and you flip them all at once. Each coin will land heads up or tails up, each with a 50 percent chance (assuming you are using a fair coin). What do you expect if you flip four coins at the same time? You expect, on average, two heads and two tails. However, by chance you might get another combination, such as three heads and one tail, one head and three tails, or even all heads or all tails. Using standard statistical methods, we could figure out the probability of any of these things happening:

4 heads and 0 tails = $1/16$
3 heads and 1 tail = $4/16$
2 heads and 2 tails = $6/16$
1 head and 3 tails = $4/16$
0 heads and 4 tails = $1/16$

Each time you flip four coins, it is possible to get a different result. The outcome depends on chance, and you cannot predict exactly what will happen but only the relative probability of what might happen.

A similar chance mechanism operates during evolution. For any given gene, there may be different forms, known as *alleles*. Evolutionary theory deals with changes in the frequency of these alleles over time. Imagine, for example, a gene that has two alleles in a population—*A* and *B*. Imagine that 50 percent of the genes in this population are the *A* allele and 50 percent are the *B* allele. If there is no mutation or selection or gene flow, we might expect that the next generation will have the same mix of alleles—50 percent *A* and 50 percent *B*. Because of chance, however, this might not happen. The frequency of *A* might increase or decrease by chance, just as we can get variation in the number of heads and tails by flipping coins. The next generation might have 58 percent *A* or 37 percent *A* or some other value, a change due to random chance.

Genetic drift happens each generation. The relative proportion of alleles can change in any direction. It could go up, then down, and then up again. Or, it could go down for several generations in a row and then go back up by chance. Given enough time, an allele can become fixed (100 percent) by chance, just as you will eventually get all heads if you continue to flip four coins repeatedly. Also, an allele can become extinct (0 percent) for the same reason. When allele A becomes fixed, allele B becomes extinct, and vice versa. The point here is that genetic drift causes a reduction in diversity because alleles are fixed or lost over time.

What does this have to do with population size? Quite simply, a basic rule of probability is that chance events are more likely when the number of events is small and less likely when the number of events is large. To understand this, let's return to the coin-flipping example. If you flip four coins and get all heads, it is not that unlikely an event—we expect to see that 1 out of 16 times. However, if you flip ten coins, the chance of getting all heads is much, much smaller; this will happen only once in 1,024 times. If you flip thirty coins, the probability of getting all heads is virtually zero (only once in 1,073,741,824 times). The same principle applies to genetic drift. Smaller populations are expected to drift more than larger populations. Because of genetic drift, smaller populations will experience more fixation and extinction of alleles and thus will have less genetic diversity than larger populations.

How does this relate to our observation that genetic diversity is larger in sub-Saharan African populations? It means that a possible reason for this regional difference is a difference in average population size. If the human population in Africa were larger than those in other regions throughout most of human evolution, then it should show greater diversity. Perhaps the regional differences in genetic diversity we see in the world today are simply a reflection of ancient differences in population size.

At first glance, we might tend to reject this idea based on the current distribution of population on our planet. After all, the most populous continent in the world today is Asia, with more than 3.6 billion people compared to almost 800 million in Africa. Shouldn't Asia show the most genetic diversity? The problem with this argument is that the above figures refer to present-day populations, which are quite different from those in the past. Present-day figures are strongly affected by the past 12,000 years of agriculture, population movements, and other cultural changes. In ancient times, when all of our ancestors lived as hunters and gatherers, the situation was likely to have been quite different. For one thing, the amount of *usable* land mass was probably lower in many parts of Europe and Asia during various ice ages. Archaeologist Fekri Hassan has estimated that more of Africa was usable than other continents and that the subtropical savanna characteristic of much of sub-Saharan Africa could have supported many more people than other climates.[23] Combined with the rather large landmass of Africa, it seems likely that more people lived in Africa than elsewhere until very recently.

I first became interested in the problem of regional differences in population size in the early 1990s. At the time, my research did not deal with the question of modern human origins. I was playing (literally) with some global data on cranial measurements across the world to test some software I had written to analyze physical measurement data. I noticed that there was a definite geographic pattern: The average amount of physical diversity in these measures was highest in sub-Saharan African populations. At the time, I was aware of the arguments based on mitochondrial DNA but had not thought about them in any detail. My observations on cranial measures fit the same pattern. My first interpretation was that this was further proof of an African replacement, but I was less sure of the underlying dynamics since the evolution of physical measures, affected by both genetics and environment, can be rather complex. Before I could give much thought to this, other projects and duties intervened, and I wound up filing all the analyses away for more than a year until I had time to reexamine them (also, I often find that time away from a problem helps in solving it).

I then started discussing these results with my colleague Henry Harpending. We started with the idea that regional differences *might* be a reflection of differences in the time depth of accumulated mutations, but that differences in population size were a more likely explanation. We also felt that regardless of the true origin of modern humans, we had to factor migration into any analytic model. Even if all modern humans arose in Africa 150,000 years ago, they had nonetheless been exchanging genes across the world since that time. We applied a method of comparing observed variation with the amount of variation expected under a model of genetic exchange between regions. Our results were again consistent with the studies of mitochondrial DNA; Africa showed considerably more diversity than expected. The only explanation allowed by the mathematics of our model was that the average population size of Africa had been larger than those of other regions over much of human evolution.

It was then that I experienced one of those rare occasions when an answer becomes immediately clear. If I kept reanalyzing the data under different scenarios of population size differences, I could find the combination that best replicated our observations. For example, what would the results look like if the African population were 1.5 times as large as Europe and Asia? Twice as large? Three times as large? Realizing this would take me a very long time to run through many possibilities, I wrote a computer program to automate the process. Several thousand analyses later, we found that our data aligned most closely with the case where Africa had three times the population size of any other region. We did not view this estimate as particularly useful by itself, but we emphasized that the observed pattern of regional differences in diversity could easily be explained by regional differences in population size.[24] Ecological and archaeological inferences also support a larger African population.[25] Although our results did not reject an African replacement model, they did not require one. Larger African population size, and consequently greater African genetic and physical diversity, could be explained under both origin models. Several years later, I conducted a similar analysis using DNA sequence data collected by my colleague Lynn Jorde at the University of Utah. We found the same results: The best fit for the data was when more ancestors lived in Africa than in any other geographic region.[26]

How Many Ancestors?

When Henry Harpending and I began working together, we favored the African replacement model. We felt that the age of "Eve" and other common ancestors didn't actually resolve the matter, nor did geographic differences in genetic diversity. Both observations could be explained by a number of different scenarios and did not prove African replacement. We did feel, however, that another source of genetic evidence provided the best support for an African replacement: estimates of the total number of ancestors who lived within the past few hundred thousand years.

How can we determine how many people lived in ancient times? Archaeologists can provide rough estimates based on total land area and likely population densities for hunting and gathering groups. Fekri Hassan has compiled such estimates and suggested a total human population of about 1 million people 200,000 years ago.[27] Estimates of population size can also be made using genetic data. We've seen that genetic diversity is related to population size. This relationship can be used to estimate past population size. There are several different ways to do this; I won't deal here with the specifics but will focus on the results. Many different studies, using several different methods, have come up with the same genetic estimate: a population size of less than 10,000 adults 200,000 years ago.[28]

The genetic estimate refers to the number of reproducing adults, whereas the archaeological estimate refers to the *total* number of people, young and old. Given that roughly one-half to one-third of a population is generally composed of individuals of reproductive age, the genetic estimate of 10,000 adults translates to a total population size of about 20,000 to 30,000 people. This estimate is *much* lower than the archaeological estimate of 1 million!

Why are these estimates so different? Perhaps not all of the humans who lived in the past are our ancestors. This difference is easily accommodated by the African replacement model, which proposes that the non-African archaics became extinct. They lived and died but did not contribute to our genetic ancestry. They would be picked up in any archaeological analysis but would not be counted in a genetic analysis because they left no genes. The genetic estimates pick up only those who actually contributed genes to future generations.

The small number of ancestors suggested by genetic analysis implies that our ancestors lived in a single place, such as Africa. Given typical population densities for hunting and gathering populations, a figure of 20,000 to 30,000 people corresponds to a region roughly twice the size of California. This is compatible with the African replacement model; modern humans arose as a new species in a small location in Africa, later spreading out and growing in number. At first glance, the small population size does *not* seem compatible with multiregional evolution; if our ancestors were so few in number, it seems unlikely that they could spread out over several continents and remain connected. If you took 25,000 people and distributed them over three continents in small local groups, they would be so widely scattered that they would very rarely meet or exchange genes, something required under multiregional evolution.

This is essentially the view that Henry Harpending and I held at the time we wrote several papers about modern human origins. To me, the conclusions of the studies conducted by Cann and her colleagues (and others) did not solve the problem, but the estimates of the human ancestral population size did. In other words, they were right, but for the wrong reason. As time went on, however, I began to question our original interpretation. It turns out that genetic estimates of total population size are not always accurate; these estimates are influenced by factors such as variation in fertility and fluctuations in population size over time. In particular, the dynamics of local groups greatly affects the genetic estimates. If, as is common in many organisms, there is frequent extinction of small groups combined with recolonization from genetically related neighboring populations, then genetic estimates of population size will be too low.[29] My preliminary work in this area suggests that the genetic estimates we see are easily compatible with a total population of several hundred thousand or more humans. Thus, as with other genetic data, estimates of population size based on genetic diversity are compatible with both African replacement and multiregional evolution models of modern human origins.[30] I now think that my initial support of an African replacement model based on small population size was in error. The genetic estimates tell us something about ancient population dynamics but do not necessarily provide an accurate count of past population size.

Mostly Out of Africa?

In some ways, my own research on modern human origins has been very frustrating, because I would like to find the one piece of evidence that supports one model to the exclusion of the other. Much evidence that at first appears to support the African replacement model winds up also being easily explained under a multiregional model. Contrary to what is often claimed in both the scientific and popular literature, the debate over human origins continues.

Genetic analysis alone does not provide a definitive answer. I do feel, however, that the genetic data combined with observations from the fossil record do give us a picture of a likely model of modern human origins. The fossil record, although not complete, indicates that the first changes to modern human anatomy took place in sub-Saharan Africa, at least 130,000 years ago. These changes spread outward from Africa over time, appearing next in the Middle East, and then later in Asia and Australia, and finally in Europe. The genetic data are consistent with this. What is less clear, however, is what happened outside of Africa, where we know that more archaic humans were living. Were these archaics replaced? I tend to doubt it. I suspect that what actually occurred was a mixture of populations and genes over time and that this mixture was strongly affected by differences in ancient population size.

Imagine populations across the Old World interconnected by gene flow across both short and long distances. In any generation, most individuals choose mates from their local population, but there is always a trickle of gene flow connecting it to other populations both close by and far away. Over time, this small amount of gene flow adds up to make populations more similar to each other. As an analogy, consider taking two gallon cans of paint, one with red paint and one with white. Take a teaspoon of red paint and pour it into the can of white paint and at the same time add a teaspoon of white paint to the can of red paint. A teaspoon won't make much difference, but if you continue doing this, you will see that the can of white paint becomes redder and the can of red paint becomes whiter, until both cans eventually become the same shade of pink.

Now, imagine the same process with one critical difference; instead of mixing two gallon cans of paint, mix a gallon can of red paint and a quart can of white paint. Now when you mix a teaspoon of paint from each can, the impact of the red paint on the smaller can of white paint will be greater than the impact of the white paint on the larger can of red paint. Eventually, both cans will reach the same color, but it will be a much redder shade of pink because you started with more red paint than white paint.

Of course, genes are not paint, but the basic idea is much the same. If Africa's population were larger than the populations of other geographic regions, then the genes flowing out of Africa would have had a greater impact on the evolution of humans than those from other regions. According to this model, the smaller populations outside of Africa would change over time to become more similar to the larger African population (except for factors that counter this trend, such as genetic drift). This process would be enhanced if there were an advantage to alleles arising in Africa; in this case, the spread of genes outward from Africa would be faster and more complete.

Thus, my own views do agree with the African replacement model to the extent that I suggest the *initial* changes leading from archaic to modern human *did* take place first in Africa and then spread outward over time. However, I think these changes took place within a single evolving species and did not involve complete, wholesale replacement. One could argue that this is a semantic difference, because what I have described could easily be taken as a form of "genetic replacement" over time. If the final can of paint is reddish, does it matter if one poured out the original can of white paint and replaced it with red, or if one simply added more and more red paint to the can of white paint over time? In terms of mixing paint, it doesn't make a difference, but the distinction is critical in understanding our own origins. The difference between the birth of a new

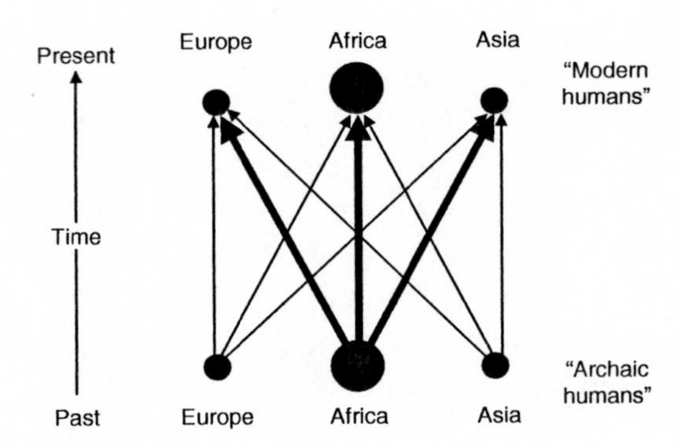

Figure 3.8 "Mostly out of Africa." My interpretation of the multiregional model has most of human evolution taking place in Africa. According to this model, most of our ancestors 150,000 years ago lived in Africa, but not all of them.

species (African replacement model) and change within a species (multiregional evolution model) is fundamentally crucial in evolutionary terms. When we look at the fossils of non-African archaic humans, we want to know whether they are an inherent part of our ancestry or a side branch having no direct kinship with us. Even a small genetic contribution would be significant. Coming back to the basic question of this chapter, I suggest that 150,000 years ago *most* of our ancestors lived in Africa, but not all of them (Figure 3.8).[31] I think that the evidence points to *some* ancient non-African ancestry, although it is not clear what was contributed by specific populations from geographic regions outside of Africa. This brings us to the subject of the next chapter. What happened to the Neandertals in Europe and the Middle East?

FOUR

The Fate of the Neandertals

What happened to the Neandertals? Where did they go?

This question is perhaps one of the most commonly asked questions about human evolution. As with many questions about disappearances, it conjures up an air of mystery. What happened to the lost colony of Roanoke, Virginia? Whatever happened to Amelia Earhart, Judge Crater, or Jimmy Hoffa? Such questions are common throughout history, and their perpetuation in our popular culture reflects, I think, a fascination with unsolved mysteries. The fate of the Neandertals is similarly mysterious to many. As described in the previous chapter, the Neandertals were a group of large-brained archaic humans with several distinctive physical characteristics, such as their large noses and faces. They lived in Europe and the Middle East (Figure 4.1) but are now no longer with us. I think that the fate of the Neandertals is a particularly interesting mystery because it deals with an entire group of people, not just a single individual or colony. This mystery is compounded by their simultaneous similarity and dissimilarity to us, being like us in many respects, but different in others. If they died out, then why did we survive?

In our popular culture, the fascination with Neandertals and their fate is often confounded by continued misunderstanding about who they were and how they lived. Even the name is confusing. The name Neandertal translates from German as "Neander Valley" (*tal* means "valley" in German), named after the location of the Feldhofer Cave site in Germany where the first specimen was discovered in 1856. An alternative spelling is often used that includes the silent letter *h* (Neanderthal). Scientifically, Neandertals have been given different names as well. William King proposed the formal species name of *Homo neanderthalensis* in 1864 to emphasize his interpretation of their difference from living humans.[1] By the

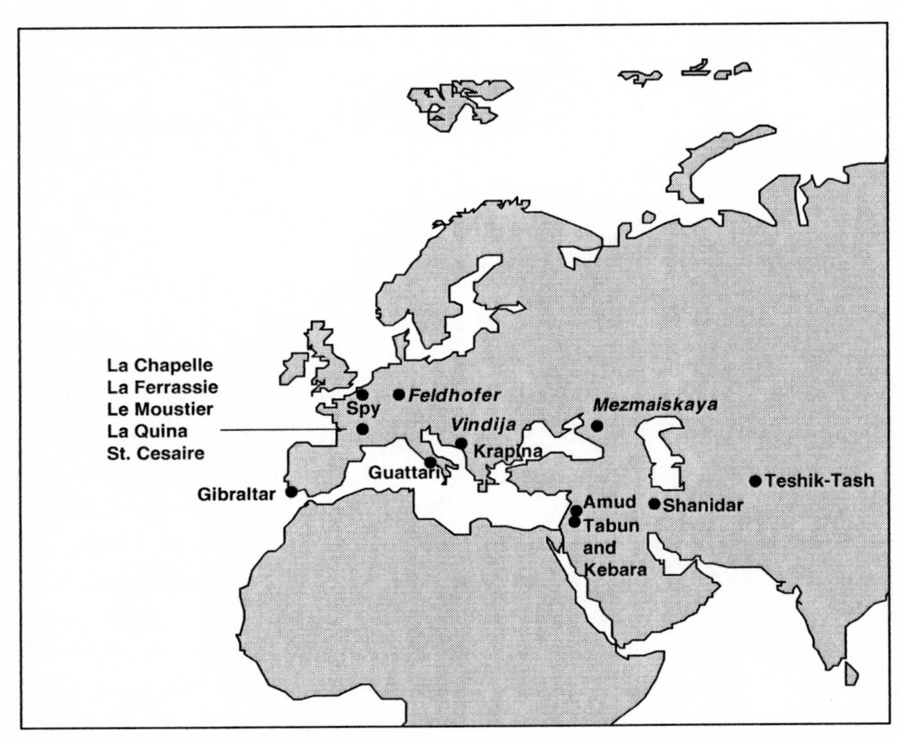

Figure 4.1 The geographic distribution of some Neandertal sites in Europe and the Middle East. The three locations indicated in italics (Feldhofer, Mezmaiskaya, and Vindija) are the sites from which mitochondrial DNA has been extracted from fossil specimens.

middle of the twentieth century, most anthropologists viewed the Neandertals as a particular variant of *Homo sapiens* and frequently referred to the group as a unique subspecies—*Homo sapiens neanderthalensis*—to distinguish it from living humans, who are scientifically referred to as the subspecies *Homo sapiens sapiens.* The subspecies designation was a useful way of acknowledging the Neandertals' similarity to living humans (same species and therefore capable of exchanging genes with other humans) while emphasizing their differences (a distinct subspecies with unique traits). By the end of the twentieth century, some felt that the differences between Neandertals and living humans were too great to lump both groups under the same species, and the species name *Homo neanderthalensis* regained popularity in some circles. The debate over whether Neandertals were a separate species continues today.[2]

The Neandertals (regardless of their evolutionary relationship to us) have endured considerable bad press. The name has become a common insult, frequently heard in situation comedies to refer to someone who is a bit slow or brutish. Such insults play into an image from popular culture that Neandertals were not terribly bright, could barely cope with day-to-day problems, did not walk fully upright, and were semi-human savages. Although false, these images persist to this day. Many of these ideas stemmed from an inaccurate reconstruction of a Neandertal skeleton in the early twentieth century. One specimen had curved thigh bones, which were interpreted as a sign that Neandertals walked in a bent-over posture. In actuality, this individual was an old man who had had severe arthritis. Neandertals were in fact fully bipedal.

Contrary to popular culture, Neandertals were not idiots. They had large brains and were intelligent. They made sophisticated stone tools that required considerable thought and insight, hunted large animals, controlled fire, and buried their dead in caves. They lived and survived under extremely harsh conditions in glacial Europe. These were not the Stone Age buffoons portrayed in a number of bad movies!

Although some Neandertal features are still found among living humans, no one has the complete mix. As a distinct group, they no longer exist, which brings us back to the basic question of their fate. Did they become extinct? If so, how? Did they change over time, eventually evolving into modern Europeans? Did they interbreed with an incoming wave of modern humans out of Africa and become genetically absorbed into a larger gene pool in the process? Many different ideas have been proposed over time to answer such questions. For our purpose here, the major question is whether they left any genes behind in us. Does *part* of our ancestry trace back to the Neandertals, or did they *completely* vanish leaving only their bones and artifacts behind?

The question of the fate of the Neandertals is tied closely to the debate over modern human origins discussed in the previous chapter. Under the African replacement model, the Neandertals were one of a number of populations of archaic humans that became extinct when modern humans dispersed out of Africa beginning 100,000 years ago. Their fate here is clear—they died out—although there is debate about how that actually might have happened. The fate of the Neandertals is less clear when considering multiregional evolution in the broadest sense, which proposes that *some* (not necessarily all) archaic populations outside of Africa contributed to the ancestry of living humans. In some versions of multiregional evolution, the Neandertals are considered *part* of our ancestry. It is also possible to envision a multiregional model where *some* non-African populations are part of our ancestry, but Neandertals are not.

The Fossil Record of Neandertals

What do we actually know about our ancestral connection with the Neandertals? By the start of the twentieth century, the fossil record in Europe showed a noticeable gap between Neandertals, who appeared to have lived until about 35,000 years ago, and the first European modern humans, who appeared to have arrived on the scene about 30,000 years ago. Much of the early debate over Neandertals focused on the meaning of this 5,000-year gap. Was there an evolutionary connection between the Neandertals and moderns? Was 5,000 years long enough in evolutionary time to change one into the other? Arguments flew back and forth between those championing a replacement scenario and those who considered the Neandertals as a "phase" between earlier humans (such as *Homo erectus*) and modern humans.

Continued discoveries showed that the problem was even more complex. The discoveries of Neandertals from the Middle East showed a greater geographic range than first imagined. The Middle Eastern Neandertals were considered by some to be more similar to modern humans. There appeared to be a trend in the Middle East that showed earlier Neandertals giving rise to more modern-looking humans. Anthropologist F. Clark Howell suggested that the Middle Eastern Neandertals gave rise to both modern humans and the "classic" European Neandertals, who subsequently became more anatomically specialized, perhaps because of adaptation to a cold glacial environment.[3] By the 1960s, an increasing number of anthropologists viewed the Neandertals as a regional variant of an evolving line of humans leading to modern *Homo sapiens.*

By the late twentieth century, it became apparent from new geological dates that the idea of a simple transition from Neandertals to modern humans was incorrect. The fossil evidence from Europe now shows a period of overlap in time for Neandertals and "moderns." The youngest Neandertals are now dated at 28,000 years, showing coexistence with modern human populations in parts of Western Europe for at least several thousand years.[4] Neandertals and moderns lived near each other for a time. Did they interact? If so, how? Some have suggested there was some cultural contact with the Neandertals adopting certain aspects of stone tool technology and material culture from the moderns. From our perspective, the relevant question is whether they mated and produced fertile offspring. If the answer is yes, then we should consider Neandertals and moderns as belonging to the same species. If the answer is no, we must then consider a situation of two separate but related species, both culturally adapted, living side by side until one died out, for whatever reason.

As noted in the previous chapter, it is not as easy as it might sound to answer questions about species from the fossil data. Some anthropologists see a number of distinct anatomic features that suggest a long period of genetic isolation that led to separate species. Others see evidence of Neandertal traits in the first post-Neandertal European modern humans (although at reduced frequency), suggesting some genetic continuity (and thus interbreeding) over time. More recently, the fossil of a Neandertal child from the Lagar Velho site in Portugal has fueled further debate. This fossil, described by Cidália Duarte and her colleagues, is that of a four-year-old child dating to 24,500 years ago that has a mixture of Neandertal and modern characteristics.[5] Certain features, such as the teeth and skull, show affinity with moderns, whereas other features, such as body proportions and certain muscle insertions, more closely resemble those of Neandertals. Although the authors interpret this mosaic of modern and Neandertal traits as a signal of past mixture between the local Neandertal populations and early modern humans moving into the region, others have suggested that the child is simply a rugged-looking modern human and that there is no evidence of genetic mixture.[6]

The situation for the Middle Eastern Neandertals has also become more complicated, with some arguing for a long period of coexistence between Neandertals and moderns. Geologic dates now show that some early modern forms lived *before* the Middle Eastern Neandertals, thus ruling out any simple model of Neandertals evolving directly into moderns.[7] Others question whether the two groups actually lived in the Middle East at the same time, since Neandertal and modern fossils are separated by thousands of years. Some have suggested that the region might have been inhabited by both groups but at different times, in a prehistoric version of time-sharing. According to this view, both modern and Neandertal populations expanded and contracted their geographic range because of climatic change.[8] During the coldest of times, the Neandertals from Europe moved south into the Middle East, while the moderns moved back into Africa. Finally, not everyone accepts the idea that the Middle Eastern fossils represent two different types; some assert that they might comprise a variable population of early moderns and that the term Neandertal should therefore be confined specifically to the European fossil record.[9]

Although we have learned more about what Neandertals looked like, and where and how they lived, there remains the nagging question of their fate. Historically, the arguments have focused on different analyses and interpretations of the fossil record, and this is likely to continue. Earlier views of a simple transition from Neandertals to moderns are pretty much ruled out, but the question of genetic relationships remains. With this brief review of a long, and often complex, history of scientific debate, we turn to considering what genetic data can tell us about the fate of the Neandertals. Unlike the kinds of genetic analyses described in the previous chapter, which rely on inferences made based on the genetic diversity of *living* humans, the Neandertal question now has input from another kind of genetic data, one only dreamed about in science fiction until a few years ago—the analysis of actual DNA from Neandertal fossils.

The Discovery of Neandertal DNA

Readers of science fiction are familiar with plot lines that involve the reconstruction of past life from traces of ancient DNA. Such stories are now part of our popular culture, owing to the success of three *Jurassic Park* movies in which ancient dinosaur DNA is extracted from prehistoric blood-sucking insects trapped in amber. The ability to reconstruct an entire prehistoric creature is not something that we can do (at least so far), but we can extract small sequences of ancient DNA. Extracted DNA sequences represent only a tiny fraction of an organism's total DNA, so we should not expect to see the complete genetic blueprint for a Neandertal nor the ability to produce a Neandertal in the lab (outside of science fiction, don't expect to see a real-life *Pleistocene Park*). However, the small fraction of DNA that can be extracted from ancient organisms does provide us with clues to evolutionary relationships.

How can we obtain ancient DNA? A major breakthrough came with the invention of a molecular laboratory technique known as the polymerase chain reaction (PCR), a method developed by Kary Mullis in the 1980s, for which he was awarded the Nobel Prize in chemistry in 1993. PCR is a method that allows the amplification of small amounts of DNA. In simple terms, DNA is used as a template for duplicating more DNA, using the molecule's ability to make copies of itself. The amplified DNA is then put through the process repeatedly, each time doubling the amount of DNA, until the small amount of initial DNA has been amplified considerably. Thus, one molecule of DNA will yield two molecules after the first cycle, four after the second, eight after the third, and 1,048,576 molecules after twenty cycles.

The ability to make a large sample of DNA from a small initial amount has transformed molecular biology, with applications in everything from forensic applications (such as DNA fingerprints) to reconstruction of ancient DNA. We now have a tool that allows us, under certain conditions, to probe directly the genetic code of past humans. However, the method is not quite as simple as it might sound; it does not consist of placing a fossil into a test tube, pushing a button, and getting a complete readout of the DNA sequence. For one thing, DNA degrades over time, and there may not be enough left to replicate after thousands of years of decay. Particular conditions, such as the inside of caves, are required for the preservation of DNA, and therefore the methods may not always be successful in all parts of the world. In addition, much ancient DNA research is limited to mitochondrial DNA because there are more copies of this in a cell than nuclear DNA. Thus, we are more likely to obtain mitochondrial DNA.

Another major problem is contamination. PCR is so sensitive that it can often replicate small amounts of DNA from skin cells floating in the air. If you handle a fossil specimen, some of your DNA is rubbed off, and you have to be careful that your analysis is actually detecting the fossil's DNA and not your own. Imagine, for example, that you extracted DNA from a Neandertal fossil and found that it was identical to many living humans. You might conclude that there is no genetic difference between Neandertals and modern humans, but first you ought to check to see whether the DNA you analyzed is your own!

There are, of course, different lab protocols and preliminary tests that can address the above issues. It is also best to run the analysis at more than one lab to confirm results. Another significant problem is that to obtain ancient DNA, some of the original fossil must be destroyed. Because fossils are irreplaceable treasures that cannot be duplicated, the potential benefits of ancient DNA analysis have to be weighed against the partial destruction of the fossils.

In 1997, Matthias Krings of the University of Munich, along with colleagues in Germany and the United States, announced the successful extraction of a mitochondrial DNA sequence from the Feldhofer Cave Neandertal.[10] A small amount of fossilized bone was removed from this specimen's upper right arm for DNA analysis. Initial protein analysis suggested that this specimen contained amplifiable DNA (if it didn't, there would be no point in continuing). They then extracted mitochondrial DNA from this specimen and performed a number of tests to ensure that the resulting sequences were not contaminated, including cross-checking in two independent laboratories.

Krings and his colleagues were able to obtain a sequence of 379 DNA positions (known as base pairs, and abbreviated as "bp") from a section of mitochondrial DNA known as hypervariable region 1. This region accounts for only a small fraction of the total mitochondrial genome (which is over 16,000 bp in length) but is particularly informative when making evolutionary comparisons. They then compared the Neandertal DNA sequence with sequences from living humans, starting with a comparison with what is known as the Human Reference Sequence. When comparing DNA sequences, we look for mismatches—cases where the DNA base (A, T, C, or G) is different. They found twenty-seven differences between the Feldhofer DNA sequence and the Human Reference Sequence.

Because not all living humans have the exact same mitochondrial DNA sequence as the reference sequence, it is also useful to compare the Neandertal sequence with sequences from a number of different living humans. Krings and his colleagues did this by taking 994 mtDNA sequences from around the world and comparing each of them to the Neandertal sequences. As expected, some were more (or less) similar to the Neandertal sequence than others. They found that the *average* number of differences was 27.2. Is this a large genetic difference or a small genetic difference? One way of addressing this question is to look at the average number of differences between living humans, which was done by comparing each of the 994 DNA sequences from living humans to all others—a total of 493,521 comparisons. The average number of differences between living humans is 8.0. Therefore, the average difference between the Feldhofer Neandertal and living humans (27.2) is more than three times the average difference between living humans (8.0).

It is also useful to examine the range of differences. When comparing living humans to Neandertals, the number of differences ranged from a low of 22 to a high of 36. When comparing living humans to each other, the number of differences ranged from a low of 1 to a high of 24. There is some overlap here, as there are a *few* cases where the difference in mitochondrial DNA between living humans is greater than the number of differences between the Feldhofer Neandertal and living humans. However, the degree of overlap is very small—only 0.002 percent.

An additional set of comparisons compared the Feldhofer sequence to sequences obtained from living chimpanzees based on 333 base pairs (as compared with 379 bp compared above; the number of comparisons depends on how much of an mtDNA sequence is available in common for different samples). They found that the average number of differences among living humans for this length of DNA was 8.0, compared to an average of 25.6 differences out of 333 basic pairs, separating the Feldhofer sequence from living humans and an average of 55.0 differences between living humans and chimpanzees. These analyses show that the average genetic difference between living humans and the Feldhofer Neandertal was about half of that between chimpanzees and living humans ($^{25.6}/_{55.0} = 0.47$).

Krings and his colleagues interpreted the large genetic difference between Neandertals and living humans as being most consistent with the idea that the Neandertals were a side branch of human evolution that most likely did not contribute to our ancestry. That is, Neandertals were a different species. Further, they constructed a gene tree and estimated that the most recent common female ancestor shared by Neandertals and living humans lived between 550,000 and 690,000 years ago. Two years later, Krings and his colleagues published a paper presenting the results of further extraction of mitochondrial DNA from the Feldhofer Neandertal, this time based on another part of the mitochondrial DNA sequence.[11] These results basically confirmed the earlier analysis; Neandertals were genetically different from living humans.

A number of anthropologists took these results as the final answer to the question of the fate of the Neandertals. It appeared to many that the Neandertals were *not* part of our ancestry but instead were closely related cousins occupying a side branch in our family tree. Modern humans share some kinship with them, but the two groups have been separate evolutionary lines for more than half a million years. Neandertals were replaced by a new species, *Homo sapiens,* originating in Africa and dispersing into Europe starting about 40,000 years ago. According to the replacement interpretation, Neandertals hung on for a while in isolated parts of Western Europe but eventually died out by 28,000 years ago, leaving only modern humans. Although this interpretation has been accepted by a number of scientists, not everyone agrees, as will be described shortly.

More Neandertal DNA

Even given the strict protocols of the Neandertal DNA study, some people wondered if the results might have been a fluke. Such concerns were laid to rest by two subsequent studies that extracted mitochondrial DNA from other Neandertal fossils. In 2000, Igor Ovchinnikov and his colleagues reported the extraction of mtDNA from a Neandertal infant dating to 29,000 years ago from Mezmaiskaya Cave in the northern Caucasus region of Europe.[12] Again, preliminary tests showed adequate molecular preservation of DNA, allowing them to extract a 345-base-pair sequence of mitochondrial DNA from two rib fragments. When the Mezmaiskaya Cave fossil sequences were compared to the Human Reference Sequence, researchers found a total of 23 differences.[13] When the same sequences were compared to the mtDNA of the Feldhofer Neandertal, 12 differences were found. Later in 2000, Krings and his colleagues published an analysis of mitochondrial DNA obtained from a third Neandertal specimen, this one from Vindija Cave in Croatia, dating back about 42,000 years.[14] Once again, the results showed the Neandertal sequence to be rather different from that of living humans. The question is *how* different?

Taken as a group, the three Neandertal sequences show a number of differences from the Human Reference Sequence (Figure 4.2). Although there is variation among the three Neandertals (which were separated in both time and space), the DNA sequences all tend to be more similar to each other than any are to living humans (Figure 4.3).

Neandertals: Different Species or Different Subspecies?

At first glance, the large difference between the mitochondrial DNA of Neandertals and that of living humans seems to support the view that the Neandertals were a separate species from us, a side branch of human evolution that eventually became extinct. It is certainly the case that Neandertal mtDNA *is* different on average from that of living humans, but

```
Positions 16056-16115

Human Reference Sequence  CCAAGTATTGACTCACCCATCAACAACCGCTATGTATTTCGTACATTACTGCCAGCCACC
Feldhofer                 ......................G..............C.............TT..TT...
Mezmaiskaya               ..............................C.............................
Vindija                   ......................G.....................................

Positions 16116-16175

Human Reference Sequence  ATGAATATTGTACGGTACCATAAATACTTGACCACCTGTAGTACATAAAAACCCAATCCA
Feldhofer                 .............A.........T........T.....C..............T......
Mezmaiskaya               .............A.........T........T.......A............T......
Vindija                   .............A.........T........T.....C..............T......

Positions 16176-16235

Human Reference Sequence  CATCAAAACCCCCTCCCCATGCTTACAAGCAAGTACAGCAATCAACCCTCAACTATCACA
Feldhofer                 .......C.....C...................C.............T......G...T.
Mezmaiskaya               ......CC.....C...................C.............T......G...T.
Vindija                   ......CC.....C...................C.............T......G...T.

Positions 16236-16294

Human Reference Sequence  CATCAACTGCAACTCCAAAGCCACCCCT CACCCACTAGGATACCAACAAACCTACCCAC
Feldhofer                 ........A...........A.G...T.A..............T................
Mezmaiskaya               ........A...........A.....T.A..............T................
Vindija                   ........A...........A.G...T.A..............T................

Positions 16295-16354

Human Reference Sequence  CCTTAACAGTACATAGTACATAAAGCCATTTACCGTACATAGCACATTACAGTCAAATCC
Feldhofer                 ....G...........C........T..................................
Mezmaiskaya               ....G...........C........T.......................T..........
Vindija                   ....G...........C........T..................................

Positions 16355-16378

Human Reference Sequence  CTTCTCGTCCCCATGGATGACCCC
Feldhofer                 .......C................
Mezmaiskaya               .......C................
Vindija                   .......C................
```

Figure 4.2 Neandertal DNA sequences. Comparison of a 324-base-pair sequence of mitochondrial DNA for the Human Reference Sequence and the three Neandertal specimens. This figure restricts comparison to that section of mtDNA in common among all four sequences. Dots in the Neandertal sequences indicate positions that are the same as the Human Reference Sequence. The three Neandertal sequences share an insertion of "A" immediately after position 16263 that is not found in the Human Reference Sequence.
Source: Krings et al. (1997, 2000); Ovchinnikov et al. (2000); and GenBank (http://www.ncbi.nlm.nih.gov/Genbank/).

does this necessarily mean Neandertals were a different species? For many decades, paleoanthropologists held that the Neandertals were not a separate species but a different *subspecies*, based on their anatomical similarities to modern humans.

One way of answering this question is to consider how much variation in mitochondrial DNA is likely to exist among subspecies within a single

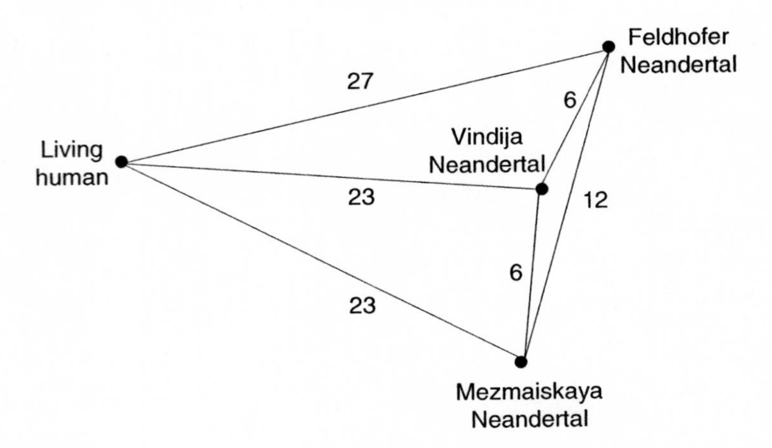

Figure 4.3 DNA differences between living humans and Neandertals. This graph is a schematic diagram showing the number of mitochondrial DNA differences between the Human Reference Sequence and the three Neandertal sequences. These comparisons are based on the 324-base-pair sequence common to all samples. Neandertals, as a group, are different from living humans. Data from Figure 4.2.

species. One comparison is with chimpanzees, who along with bonobos are our closest living relatives. There are three different subspecies of chimpanzee in the world today. These different populations are labeled as subspecies because they all belong to the same species *(Pan troglodytes)* but are geographically and biologically rather distinct from one another. By comparing the mtDNA variation among chimpanzee subspecies, we get an idea of how much variation to expect among subspecies within a single species. If, as has long been claimed by a number of anthropologists, the Neandertals were a separate subspecies, rather than a separate species, then a comparison of mtDNA differences between Neandertals and living humans relative to variation among chimpanzee subspecies might give us some insight. If the number of mtDNA differences between Neandertals and living humans is *greater than* that found between chimpanzee subspecies, the case for separate species status for Neandertals is strengthened. However, if the number of mtDNA differences between Neandertals and living humans is *less than* that found between chimpanzee subspecies, then we cannot rule out the possibility that Neandertals were a subspecies of *Homo sapiens.* The comparison would not answer the question conclusively, but the results would lend weight to one hypothesis over the other.

Matthias Krings and his colleagues examined a 312-base-pair sequence of mitochondrial DNA in the three subspecies of chimpanzee: the central chimpanzee subspecies *(Pan troglodytes troglodytes),* the western chimpanzee subspecies *(Pan troglodytes verus),* and the eastern chimpanzee subspecies *(Pan troglodytes schweinfurthii).*[15] When they compared the mitochondrial DNA sequence of the three subspecies to each other, they found the following number of differences:

Central versus western subspecies = 36.2
Western versus eastern subspecies = 33.0
Central versus eastern subspecies = 19.7

For this same length of DNA sequence, the average difference between the Feldhofer Neandertal specimen and living humans was 25.6. The difference between Neandertals and living humans is actually *less* than the differences found in two out of three comparisons of chimpanzee subspecies (Figure 4.4).

Although these comparisons do not settle the debate, they do offer evidence that the difference between Neandertal and living human DNA is not so large as to rule out the possibility that Neandertals were actually a *subspecies* within an evolving human lineage. Furthermore, we need to consider that the comparisons between chimpanzee subspecies involve *living* populations, whereas the comparison between the Feldhofer Neandertal and living humans spans tens of thousands of years. Given that mitochondrial DNA mutates over time, some of the differences between Neandertals and living humans might simply reflect the passage of time. Even given this likely inflation, the differences between Neandertals and living humans fit comfortably within suggested limits of differences between subspecies of a single species. Again, this does not demonstrate that Neandertals were necessarily a subspecies rather than a separate species, but that they *could* have been and thus were capable of exchanging genes with modern humans. The observed DNA differences between Neandertals and living humans does not rule out either possibility.

Where Did All the Neandertal Sequences Go?

The comparisons described above deal only with the actual number of base-pair differences and do not consider the overall pattern of relationship shown by analyses that infer the likely path of base-pair evolution. Considering these relationships, scientists have shown that the mitochondrial

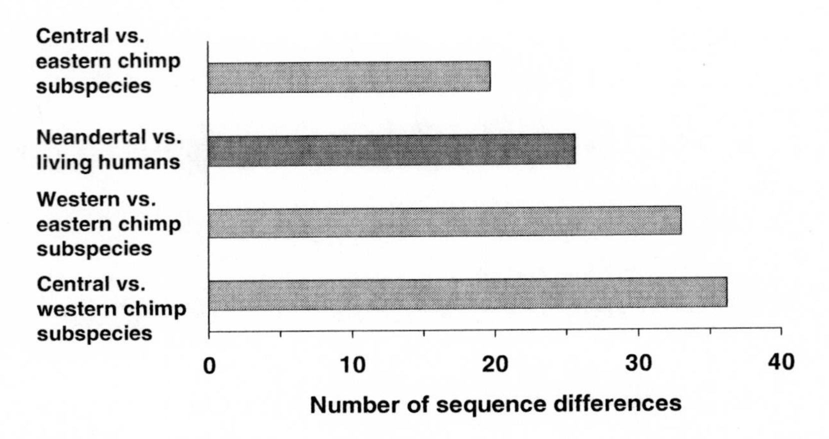

Figure 4.4 DNA differences between living humans and Neandertals compared to DNA differences among chimpanzee subspecies. This graph shows the number of mitochondrial DNA differences between Neandertals and living humans relative to the number of differences among chimpanzee subspecies based on a 312-base-pair sequence. The genetic difference between Neandertals and living humans is *less* than the difference found in two out of three comparisons of chimpanzee subspecies. These results suggest that Neandertals might have been a different *subspecies* of human.
Source: Krings et al. (1999).

DNA of Neandertals is quite different from that found in living humans. This is exactly the pattern we would expect to see if Neandertals were a separate species that branched off from our line half a million years ago or more. Could this same pattern be expected if the Neandertals were part of our ancestry? If so, why have we not found any mtDNA sequence in living humans that is as divergent as that of the Neandertals? Where did the Neandertal DNA go? The simplest explanation for not finding evidence of Neandertal mtDNA in our gene pool is that they did not contribute to it, but instead belonged to a separate and now extinct species.

This might be the case, but there is also another possibility. Perhaps the Neandertals were part of our ancestry, but their mitochondrial DNA became lost over time due to genetic drift. As described in the previous chapter, genetic drift leads to the loss of genetic variants over time. This process is particularly apparent when a trait is inherited from only one parent. As an analogy, consider the inheritance of surnames, which in many cultures are inherited through only one parent's line. Typically, sons and daughters inherit their father's last name, and daughters inherit their husband's name upon marriage. Imagine, for example, that a given family has four children. All other things being equal and assuming a 50:50 sex ratio, we expect that two out of four children will be sons and will pass along their father's surname to their children. However, this average expectation may not hold in all cases. In fact, probability theory shows that we would find two sons and two daughters only 37.5 percent of the time. In other cases, we might get by chance four sons (6.25%), three sons (25%), one son (25%), or no sons (6.25%). If this hypothetical couple has four daughters and no sons, which is expected to happen 6.25 percent of the time, then the surname becomes lost. Even though the daughters continue to pass on their genes, the surname will disappear. Historical studies of surnames show the extinction of surnames is very common after only a handful of generations. The same principle applies to genetic drift and mitochondrial DNA. If a given couple has no daughters, the mitochondrial DNA is lost. For example, my three children (all sons) carry their mother's mitochondrial DNA but cannot pass it on. Although genetic drift occurs for all genetic traits, it is even more dramatic for mitochondrial DNA.

Considering drift and DNA sequence extinction raises the question of whether it is possible for the Neandertals' mitochondrial DNA to have been lost while some of their nuclear DNA persists. Although one way of dealing with this would be to examine large sections of nuclear DNA from Neandertal fossils, nuclear DNA degrades too quickly to be reliably extracted from such ancient fossils. For the moment at least, we have to rely on the evidence of mitochondrial DNA. The problem, of course, is that if we accept mitochondrial lineage extinction due to drift as a likely possibility, then we would have two models predicting the same result—the absence of Neandertal mtDNA in living humans. It is not clear which explanation is more likely.

One thing that we would very much like to know is the mitochondrial DNA sequences of anatomically modern humans found at the same time as the Neandertals. The fossil evidence clearly shows that both Neandertals and anatomically modern humans existed in Europe at the same time for at least several thousand years. There is disagreement about whether there was any genetic connection between the two. If we had mtDNA sequences from some of the anatomically modern specimens, such as the famous fossils at the Cro-Magnon site in France, then comparisons with Neandertal and living human DNA might offer some resolution. So far, we do not have such evidence (at least at the time of this writing).

While waiting for such evidence, we can examine some possible outcomes. Suppose, for example, that the mtDNA sequence of an anatomically modern human fossil found at the same time as the Neandertals looks a lot like that of living humans but is distinct from Neandertals (and assume that we have ruled out contamination). Since we know that anatomically modern humans were part of our ancestry, such a comparison would suggest little change in our mitochondrial DNA over time, and the distinctiveness of the Neandertal mtDNA would argue strongly for Neandertals having been a separate species. On the other hand, if the anatomically modern human fossil had mtDNA similar to that of the Neandertals, then that would argue strongly for the inclusion of Neandertals in our own ancestry. Of course, the problem would be how to interpret the results if the anatomically modern human fossil DNA were different from both.

Although we do not yet have any mitochondrial DNA sequences from ancient modern humans in Europe, we do have some evidence from Australia. In 2001, Gregory Adcock and his colleagues reported on the extraction of mtDNA from ten ancient Australian fossils, all confirmed to be anatomically modern in morphology.[16] Of particular interest is the mtDNA sequence from the 60,000-year-old Lake Mungo 3 specimen. This specimen is older than at least two of the three Neandertal specimens discussed above, is anatomically modern, but has a mitochondrial DNA sequence rather different from that of living humans (a modal difference of 12 DNA substitutions). Furthermore, the Lake Mungo 3 sequence is also quite different from that of more recent fossil humans from the same region.

The Lake Mungo 3 DNA sequence provides an example of a mtDNA lineage that has become extinct over time. The fossil is anatomically modern and shows continuity with later humans, but its mitochondrial lineage has been lost. This means the same thing *could* have happened with the Neandertals. It is therefore possible that the Neandertals were part of our ancestry but that their more divergent mitochondrial DNA became lost over the course of many generations.[17] This means there are two possible interpretations based on the absence of Neandertal mtDNA sequences in living humans: (1) They are not there because the Neandertals were a separate species, or (2) the Neandertals were part of our ancestry, but their mitochondrial DNA has been lost over time because of random chance. Thus, the observed data can be used to support both hypotheses of the fate of the Neandertals. The issue is clearly not resolved. In addition, the evidence from the Lake Mungo 3 specimens is still judged by some to be controversial and in need of replication.

European Affinities of Neandertal DNA

It has been suggested that if the European Neandertals were part of our ancestry, then there should be some genetic affinity with living Europeans. Several analyses have compared Neandertal mitochondrial DNA sequences to living humans in Europe, Africa, and Asia to see if there is greater similarity to living humans in one geographic region than in others. The basic premise underlying these analyses is that if the European Neandertals are part of our ancestry, then their DNA should be more similar to the DNA of living Europeans than that of living Africans or Asians.

The logic underlying this idea is as follows. According to the multiregional evolution model, human populations in the past were connected to each other through gene flow. Given the restricting effect of geographic distance on migration, the vast majority of individuals in any generation are expected to choose mates from nearby groups, with only a small proportion from farther away. Since the European Neandertals lived in Europe, this logic can be extended to suggest that over the course of many thousands of years, the majority of the gene pool of modern Europeans came from the Neandertals. If so, then when Neandertal DNA is compared to that of living humans from different geographic regions, we should see fewer differences between Neandertals and living Europeans than between Neandertals and living Africans, Neandertals and living Asians, and so forth.

If, however, Neandertals were not part of our ancestry but instead had branched off as a separate species before modern humans moved out of Africa, then we would expect to see a different pattern. Here, the number of differences between Neandertals and living humans in different geographic regions should be approximately the same because modern humans spread out into different parts of the world *after* the split of the Neandertal line from our ancestors.

The original study of the Feldhofer Neandertal by Matthias Krings and his colleagues looked at regional comparisons.[18] They found the following number of differences in 379 base pairs of the first hypervariable region of mitochondrial DNA:

Neandertal versus living Europeans = 28.2
Neandertal versus living Africans = 27.1
Neandertal versus living Asians = 27.7
Neandertal versus living Native Americans = 27.4
Neandertal versus living Australians and Oceanic peoples = 28.3

It is clear from these numbers that the genetic linkages between Neandertals and living Europeans are no greater than the genetic linkages between Neandertals and any other geographic population. The numbers are all approximately the same and are not statistically different. According to the authors, this pattern is *not* what is expected under the multiregional evolution model but is instead what is expected under a replacement model. The obvious conclusion is that the data are most compatible with a replacement model where the Neandertals were a separate species. Other analyses of Neandertal DNA show the same results for a second sequence from the Feldhofer specimen,[19] as well as the Mezmaiskaya Cave specimen.[20] These results would seem at first to provide compelling evidence that Neandertals did not contribute any mitochondrial DNA to living humans, an observation often taken as proof that they were a separate species.

The regional comparisons of DNA from living humans with Neandertal DNA fit the prediction of a replacement model. However, for this to constitute proof, we must show that a multiregional model gives a different prediction. At first glance, this seems to be the case, because if the European Neandertals contributed any genes to modern Europeans, then their DNA should be more similar to the DNA of living Europeans than to the DNA of living humans elsewhere.

Although this evidence for Neandertal extinction seemed pretty clear-cut, I began to question whether the situation was actually that simple. From my own research on human population genetics, I knew that the most obvious and intuitive answer was not always the correct one. I wondered in this case whether we should actually expect Neandertal DNA to be more similar to DNA from living Europeans than to DNA from those living in other geographic regions if genes were exchanged between regions over many generations.

To answer this question, I looked at standard methods used to determine the genetic effects of migration between a set of populations over time. Under such models, and given enough elapsed time, a balance is reached where the degree of ancestry in living groups from past groups is the same for all living groups. This probably sounds rather confusing, but a simple example illustrates this concept. Consider two initial populations A and B; in each generation, population A receives 1 percent of its genes from population B, while population B receives 4 percent of its genes from population A. For each generation, we can track the proportion of accumulated ancestry from the initial gene pools of A and B. In order to keep track of this, I'll refer to the *initial* gene pools as A and B, and use the symbols A' and B' to refer to the changing gene pools over time.

After one generation of genetic mixture the gene pool A' is made up of 99 percent A and 1 percent B, while the gene pool B' is made up of 4 percent A and 96 percent B. This follows from the amount of mixture I specified. Over time these amounts will change, reflecting the accumulated ancestry over the course of subsequent generations. These proportions can readily be figured out using a branch of mathematics known as matrix algebra, although my focus here is on the results and not the exact computational methods. After 2 generations, the genetic composition of A' and B' is

Gene pool	Accumulated ancestry from initial population A	Accumulated ancestry from initial population B
A'	98.0 %	2.0 %
B'	7.8 %	92.2 %

Nothing too surprising here—most of the ancestry in gene pool A' is still from A, and most of the ancestry in gene pool B' is still from B.

The interesting thing is that these numbers continue to change over time. For example, after 10 generations these numbers are

Gene pool	Accumulated ancestry from initial population A	Accumulated ancestry from initial population B
A'	92.0 %	8.0 %
B'	32.1 %	67.9 %

and after 50 generations the numbers are

Gene pool	Accumulated ancestry from initial population A	Accumulated ancestry from initial population B
A'	81.5 %	18.5 %
B'	73.8 %	26.2 %

At this point, *both* A' and B' have more accumulated ancestry from A than from B. This is because the rate of genetic mixture from A into B (4 percent per generation) is greater than the rate from B into A (1 percent per generation). Eventually, these changes will stabilize and result in the following genetic compositions:

Gene pool	Accumulated ancestry from initial population A	Accumulated ancestry from initial population B
A'	80 %	20 %
B'	80 %	20 %

In other words, given enough time, both A' and B' will have 80 percent ancestry relative to the initial gene pool of A and 20 percent ancestry from the initial gene pool of B. The important point here is that under such conditions, the gene pools of A' and B' will be the same. Both A' and B' will be equidistant, in terms of ancestry, from either A or B.

Now let us apply this model to the case of Neandertal DNA. If the Neandertals of Europe were part of our species and exchanged genes with other regions (in any amount) for long enough, then all living human populations, regardless of geographic region, would be equidistant genetically from the Neandertals. Thus, we would not expect Neandertal DNA to be any more similar to that of living Europeans than to that of living Africans or Asians (or other geographic populations). In other words, under certain conditions a multiregional model would produce the *same* pattern expected under the African replacement model—equidistant relationships with living humans and no particular affinity with Europe. Note the problem: Our evidence to date does not show any European affinity with Neandertal DNA, which is predicted under certain conditions by *both* multiregional and replacement models. Thus, our observation that there is no European affinity does not support one model of human origins to the exclusion of the other.[21]

Of course, the key phrase in the above statement is "under certain conditions." The genetic mixing model used above is in many ways overly simplistic, as it considers genetic change to be solely the result of gene flow. In addition, the model assumes that migration has taken place long enough to reach (or at least approach) an equilibrium state. The fact of the matter is that we really don't know enough about ancient migration rates to be able to assess this assumption more closely. For the moment, all we can do is point out that it is *possible* for a multiregional evolution model to produce the genetic patterns that we see, but we cannot prove that it did. The complication here is that we could get the same result under an African replacement model as well. In sum, the lack of affinity between Neandertal DNA and a living regional population does not at present resolve the issue. Once again, things are not quite as simple as first thought.

These simple analyses show the long-term effect of cumulative genetic exchange but run counter to our typical views on ancestry. After all, if most of someone's immediate ancestors over the past few generations came from a given part of the world, such as Africa, then it seems obvious that that person is of African descent. The small levels of gene flow over large distances would not seem at first to have much of an effect. After all, if fifteen out of sixteen of my great-great-grandparents were from Europe, then it seems reasonable to define me as being of European descent. However, over long periods of time, the cumulative effects of genetic exchange add up, and our ancestry can become quite mixed.

Where Did They Go?

We now return to the basic question posed at the start of this chapter: What happened to the Neandertals? Although it is clear that the Neandertals *as a group* no longer exist, it is less clear whether they were a separate species that became extinct, or whether some of their genes live on in our species today. Apart from the anthropological significance of the question of the Neandertals' fate, these questions relate to a broader set of concerns within evolutionary biology, namely, the pattern of long-term evolutionary change.

Modern evolutionary theory provides the tools with which to understand changes in gene pools over time. If populations within a species become isolated for a sufficient time, they can emerge on divergent evolutionary paths. Mutations that appear in one population may not appear in another. The random process of genetic drift acts differently in different populations, with the frequencies of various genes increasing by chance in some populations and decreasing by chance in others. If populations occupy different environments with different pressures on adaptation, then natural selection can act to make them genetically distinct. If, given enough time, changes in the gene pool of populations lead to enough genetic divergence so that these populations are no longer capable of mating and producing fertile offspring, then they will have become separate species. As more time elapses, continued genetic changes within each species will lead to even greater biological differences between them.

When we look at living species in the world today, their overall levels of genetic difference provide a clue to how long they have been separate, as shown in the studies described in Chapter 2 for reconstructing the evolutionary history of apes and humans. We are able to start with the fact that apes and humans constitute different species and then proceed to evaluating their genetic differences in terms of the process and duration of speciation. The situation is more difficult when dealing with the fossil record. Here, we use anatomical differences as a measure of genetic difference, and we infer that specimens belong to different species when their anatomical differences exceed what we might reasonably expect for members within a single species. We are now able to add to these anatomical comparisons the data on ancient DNA sequences and to ask whether differences in DNA sequences exceed what we would expect between members of the same species.

The history of Neandertal studies has been one of alternative interpretations of the difference between Neandertals and anatomically modern humans. Most scholars have agreed that Neandertals as a group are somewhat physically distinct from other fossil humans (although interpretation of individual specimens is often ambiguous). It is clear that the Neandertals "disappeared," but possible explanations for this evolutionary change range from complete extinction to genetic mixing.

Many have looked to genetic analysis to resolve this long-standing debate. Although much of the popular and scientific press following the first extraction of Neandertal DNA argued that genetic data *had* resolved the question and firmly demonstrated that Neandertals were a separate species from our own ancestors, I have concluded, after reviewing these arguments, that the situation is not nearly as clear as one might think. The genetic data *are* compatible with the idea of separate species status. However, they are *also* compatible with the alternative view that Neandertals were not a separate species. Thus, genetic analysis alone does not conclusively answer the question.

The history of any group of populations is likely to be marked by periods of isolation. If this isolation continues long enough, then they will become different species, so that even if they later encounter one another, they will not be able to successfully share their genes. Did this happen to the Neandertals? The fossil record at present suggests that the European Neandertals coexisted with early modern humans, presumably having moved in from the Middle East or Africa, for several thousand years. This evidence could be interpreted as follows. The Neandertals split off from other human species several hundred thousand years ago, becoming a separate human species. Around 150,000 years ago, genetic changes occurred in another group of archaic humans in Africa, leading to a new species of modern humans—*Homo sapiens.* By 90,000 years ago, some populations of modern humans had moved into the Middle East, and by 40,000 years ago, they had moved into parts of Europe. By the time the moderns and Neandertals met in Western Europe, they had already become sufficiently genetically different so that they were not able to exchange genes. Even if they mated (something we can only guess about), their offspring were not viable. Previous isolation and genetic divergence had made the Neandertals and moderns different species. Over time, the moderns replaced the Neandertals, presumably because of some basic biological or cultural advantage. The result was the extinction of the Neandertals. Both the genetic evidence and the fossil record are compatible with this view. Consider a slightly different scenario. Assume that the European Neandertals were somewhat isolated and that indeed they diverged both genetically and anatomically to some extent but that they were not isolated enough to become a different species. If so, then when the two populations met again, there could have been some genetic contact, and they would have remained within the same species.

In today's world, there are a number of cases of animal populations that, although considered to be different subspecies, are for all practical purposes different species because they do not normally have the opportunity to exchange genes. Some examples within the primates are different subspecies of gorillas and the orangutan populations living in Borneo and Sumatra. When members of these groups are artificially brought into contact by human intervention, they can still exchange genes, although actual opportunities to do so in nature are rare because of geographic and ecological isolation. The story of human evolution, however, might have been different because of our adaptive abilities, which even in the distant past allowed our ancestors to spread out over several different continents. I suggest that there were times when human populations did become isolated from one another, most likely owing to changing environmental conditions, but that this isolation was not long enough nor great enough to lead to the evolution of different species. Perhaps the Neandertals were on their way down a different road, but genetic exchange with the rest of the species, owing to the migration ability of humans, kept them from becoming a different species. This is all very speculative, but if this type of thing did happen in the past, then perhaps the fossil record in Europe is showing us a window of such contact occurring after a period of isolation. As I've argued throughout this chapter, such a scenario is compatible with the genetic evidence, but, as is the case for a replacement model, this compatibility does not prove the hypothesis.

Suppose for the moment that this model is correct, and the Neandertals mixed genetically with the modern humans moving into Europe. Wouldn't we then expect that living humans would look like a cross between an ancient Neandertal and an ancient early modern? This is not the case. Although some Neandertal traits are still found in living humans, they are low in frequency, and no one has a complete set of Neandertal features. Quite simply, they are no longer with us. Doesn't this show that the Neandertals became extinct? Perhaps, but there is another interpretation.

Recall from the previous chapter that genetic evidence suggests that throughout much of human evolution the population size of Africa was larger than any other region. If so, and if moderns came out of Africa and mixed with Neandertals, then the African population, with its greater numbers, would have contributed the larger genetic impact. If you mix 9 drops of red paint with 1 drop of white paint, you will get not a color in the middle (pink) but a solution that is mostly red. If we take this idea and extend it to models of population genetics, we would expect that the Neandertal gene pool would be smaller than the African gene pool and would tend to become even smaller over time. Under this model, the Neandertals *did* disappear as a group; they were genetically swamped by a larger gene pool. Perhaps some small portion of their genes lives on in us today but, as argued in the previous chapter, our ancestry is mostly (but not exclusively) out of Africa.

I have outlined two different scenarios (and more are possible). How can we decide between them when the genetic evidence can be read in different ways? I suggest that the fossil record of changes in Neandertal traits over time is one way to solve the problem. Several anthropologists, notably David Frayer and Milford Wolpoff, have compiled data on traits that shed some light on the fate of the Neandertals.[22] They looked at a number of anatomical traits that are considered to be either unique to Neandertals or that are found in very low frequencies in other groups. They then looked at the frequency of these traits in Neandertals, in the first post-Neandertal European moderns, and in living Europeans. As an example, consider a trait known as the suprainiac fossa, a slight depression found at the back of a skull. This trait has been found in 96 percent of Neandertal skulls, 39 percent of post-Neandertal moderns, and only 2 percent of living Europeans. If the Neandertals were completely replaced, then why would this unique trait appear in the earliest post-Nea moderns? It shouldn't be found there at all. Instead, this reduction argues for some mixing of the Neandertal and modern gene pools. However, note that by modern times, the frequency of this trait has decreased to almost zero. If the Neandertals were part of our ancestry, then they were a small part. Nonetheless, a non-zero contribution argues for a different mode of evolution than speciation and replacement.

Most other Neandertal traits show this same pattern. For example, the frequency of the horizontal-oval mandibular foramen mentioned in the previous chapter is 53 percent in Neandertals, 18 percent in post-Neandertal moderns, and 1 percent in living Europeans. These numbers suggest that Neandertal traits declined gradually over time rather than disappearing all at once, as expected under replacement. Perhaps the Neandertals did not disappear with a bang but were bred out of existence by mixing with a numerically superior gene pool.

Lest the reader think that after all the discussion of alternative interpretations presented in this chapter that I have now arrived at the point where the "truth" can be told, let me point out that the above scenario should be considered a hypothesis in need of additional testing, not the final answer. For one thing, I have glossed over many of the fine details, and I should point out that there is still debate over which anatomical features to use as reference points and which ones are present in any given specimen. Furthermore, the fossil record remains spotty, and larger samples could lead to different results. As is typical throughout science, additional data and analyses can shift the weight of evidence. Although I lean toward one particular interpretation of Neandertal history, I still regard the entire question as being unresolved. The question of the fate of the Neandertals is something we will probably be passing on to the next generation.

FIVE

The Palimpsest of the Past

At the intersection of Center and West Streets in my town there is a wide concrete wall marking one of the boundaries of Hartwick College, one of the two colleges in town. Students paint the entire wall about once a week with various messages concerning college activities. Each time they do this, the entire surface is painted over, so that when you drive by you see only the most recent message. There is no record of the past remaining for you to see. I have also seen walls that are not as completely painted over, so that you can see bits and pieces of previous messages popping out from underneath the peeling and chipping paint. Here, the image you see is a mixture of different events that have taken place at different times. What you see is a palimpsest, a document that has been partially written over so that you can see older images beneath more recent images. Specifically, the word "palimpsest" derives from the Greek word *palimpsestos,* meaning "scraped again," which referred to the practice of recycling papyrus and parchment. Old text was scraped off to use the writing material again. Not all of the older text would be erased using such methods, resulting in a document where older text could still be seen.

Genetic diversity of living humans is more like a palimpsest than a complete erasure; it reflects a mixture of past events, both recent and distant in time. Throughout written history and prehistory, human populations have moved into new territories, mated with neighbors close and far, changed in size, and undergone changes that affected genetic diversity. What we see in the world today is the combined outcomes of all such events, large and small, global and local, across the span of time. We see a composite image, one that can be analyzed by peeling away the layers of the palimpsest to reveal information on the history of human populations.

In the previous two chapters, I discussed the ongoing debate over the origin of modern humans. Both the African replacement model and the multiregional evolution model deal with the dispersal of our ancestors out of Africa into other geographic regions, though they propose differing time frames for this dispersal. As shown in the recent work of Alan Templeton,[1] it is likely that there have been several dispersals out of Africa over the past 2 million years. There is also evidence that there was migration *back* into Africa between 30,000 and 50,000 years ago, as shown in studies of Y chromosomes.[2] Regardless of whether modern humans arose as a new species or represent a mixture of a recent dispersal with previous ones, it is clear that human genetic history was not a single event but rather a continued sequence of events. Moreover, things did not stop with the evolution of the first modern humans. Archaeologically, we see evidence of a constantly changing human species. Roughly 60,000 years ago, modern humans moved to the continent of Australia. Some 15,000 (or more) years ago, modern humans moved into the Americas. Within the past several thousand years, modern humans moved farther into the Pacific, colonizing islands in Micronesia and Polynesia. Farming spread from the Middle East into Europe. Within the past 500 years, European exploration resulted in the contact and mixing of populations across the world. The advent of agriculture resulted in a rapid increase in human population size. All of these events (and many more) had an effect on genetic variation in our species.

My point here is that the patterns of genetic variation that we see in the world today were not caused by any single event, but instead reflect a palimpsest, a mosaic of events that occurred at different times and in different places. As such, our genetic diversity more closely resembles a partially painted wall, with visible images both old and new, rather than a newly painted wall where all past events have been erased. Our goal, difficult though it often is, is to peel away the different layers to determine *which* events affected current genetic variation, and *how*. Just as historians have methods to peel away the different levels of text in a palimpsest, anthropologists and geneticists have ways of doing the same thing with genetic variation.

The remainder of this book presents a number of examples of how we use genetic data on living human populations to reconstruct the history of human populations. We start in this chapter by looking at human genetic diversity from the perspective of *global* geographic patterns and the influence of history. Subsequent chapters will then look more closely at regional and local variations, attempting to link genetic variation to specific events in the past. The current chapter focuses mostly on some methods we use to reconstruct the past and presents a summary of some of the major events that will be discussed in later chapters.

Measuring Human Genetic Diversity

A variety of genetic and physical characteristics can be used to reconstruct population history. Some of these, such as mitochondrial DNA, have been described in previous chapters. When analyzing traits, it is important to focus on those that are not strongly affected by natural selection. Following the example from Chapter 3, we would not want to use skin color because it is a trait that has been shaped by natural selection, where some populations resemble each other because of parallel adaptations, not because of a genetic connection.

Anthropologists have used a wide variety of data to study population history. Early studies (before the development of the field of genetics) focused on physical characteristics, such as measures of the nose, face, and skull, among other anatomical characteristics. Although such traits are affected by environmental factors and can change during a person's life, they can be used successfully in analyses of population history with appropriate methods.[3]

Many other traits are now available to us. During the twentieth century, biochemical traits, such as blood types, were discovered, which show a more direct relationship to their underlying genetic code. Blood types are classified according to the particular molecules that are present on the surface of red blood cells, which react to specific antibodies. Most of us are familiar with the term "blood type" in the context of transfusions and blood compatibility, and with specific blood types such as "O-positive" or "B-negative." These terms are actually shorthand for *two* different blood type systems: the ABO system, controlled by a gene on chromosome number 9, and the Rhesus (or Rh) system, controlled by several genes on chromosome number 1. The ABO system has four blood types (O, A, B, and AB), and the Rhesus system has two blood types (Rh positive and Rh negative). If someone has "B-negative" blood, he has type B blood for the ABO system and type Rh negative for the Rhesus blood system.

Besides ABO and Rhesus, there are many other blood type systems, such as MN, Diego, Duffy, P, and Kell, among others. In many cases, these different blood type systems have little, if anything, to do with natural selection; they are neutral. In each case, we can estimate the frequency of the alleles, the different forms of the gene that exist in our species. For the ABO blood group system, there are three main alleles—*A, B,* and *O*—and the frequency of these particular alleles varies from population to population. In the case of the Kell blood group, there are two alleles—*K* and *k*. Vast compilations of blood group allele frequencies are available on hundreds of human populations, making them quite useful for analyzing population history.[4]

In the 1960s, scientists developed a new tool for assessing human variation from blood samples, a method known as *electrophoresis*. In this process, a blood sample is placed in a gel through which an electric current is passed. Proteins move along the path of electrical flow, but because proteins vary in their biochemical structure, some move faster than others, allowing scientists to determine which ones are present in an individual's blood. Various alleles are identified and tabulated, and genotype is established. This method has been applied to dozens of different proteins and enzymes, and as with blood types, there are large compilations of allele frequencies.

The various blood types, blood proteins, and blood enzymes (collectively referred to as *classic genetic markers*) have been very useful in the study of human population history. These genetic markers have a number of advantages over physical traits such as height or cranial shape. In addition to having a known mode of inheritance, genetic markers are constant throughout one's life. Blood type does not change with age, diet, or exercise as many physical traits do. Genetic markers allow a closer look at the underlying genetic code than physical measurements.

Since the 1980s, these classic genetic markers have lost their place as the primary source of information about human genetic diversity. Owing to new molecular genetic techniques, geneticists now spend more time studying the actual underlying DNA sequences. There are a number of ways to do this. Some DNA is studied by looking at differences in the length of DNA fragments that are cut apart by various enzymes (known as restriction fragment length polymorphisms, or RFLPs). Some DNA is studied by counting the number of copies of certain DNA sequences that are repeated (known as short tandem repeats, or STRs, and sometimes referred to as microsatellite DNA). For example, one specimen might have the sequence CACACA, which shows three repeats of the sequence CA, whereas another specimen might have the sequence CACACACACA, which contains five repeats. Finally, as discussed in earlier chapters, some analyses compare DNA sequences and count the number of differences between samples. With the recent explosion of genetic technology and the development of the Human Genome Project, geneticists now have access to the complete DNA sequence of human beings (although we are far from understanding what much of the DNA actually does). Although such advances are often thought of as having primary importance for biomedical applications, they are also a valuable resource for studying human population history.

Regardless of what specific types of data are used, an objective in many studies of human population history is to come up with a set of allele frequencies that show relationships among a set of populations. For example, imagine that we have compiled the following frequencies for a hypothetical allele for a set of four hypothetical populations:

Population 1 = 0.7
Population 2 = 0.6
Population 3 = 0.3
Population 4 = 0.1

These numbers show how frequently a given allele occurs in each population. If we have X people in a population, we have 2X alleles, since for most traits each person has two copies of a gene, one from their mother and one from their father. An allele frequency of 0.7 indicates that the proportion of genes (for a particular trait) in that population having the hypothetical allele is 0.7, or 70 percent.

Allele frequencies are used to estimate genetic distances between populations. The larger the distance, the greater the dissimilarity between populations, whereas the smaller the distance, the greater the similarity. There are many different types of genetic distance measures. Most are related mathematically to a very simple distance obtained by taking the square of the difference in allele frequencies. For the hypothetical data given above, the distance between populations 1 and 2 is $(0.7\text{-}0.6)^2$, which is equal to $(0.1)^2 = 0.01$. This number doesn't tell us much by itself. We need to look at the distance between each population and all other populations, as shown below:

Distance between populations 1 and 2 = $(0.7\text{-}0.6)^2 = 0.01$
Distance between populations 1 and 3 = $(0.7\text{-}0.3)^2 = 0.16$

Distance between populations 1 and 4 = $(0.7\text{-}0.1)^2 = 0.36$
Distance between populations 2 and 3 = $(0.6\text{-}0.3)^2 = 0.09$
Distance between populations 2 and 4 = $(0.6\text{-}0.1)^2 = 0.25$
Distance between populations 3 and 4 = $(0.3\text{-}0.1)^2 = 0.04$

What can we tell from the above numbers? Remember, the lower the distance, the more two populations are related genetically, and the higher the genetic distance, the less two populations are related genetically. In this example, the lowest distance is between populations 1 and 2, showing that they are more closely related to each other than any other pair of populations. Populations 3 and 4 also have a relatively low genetic distance, showing that they too are a closely related pair. Note, however, that all of the other distances are much larger, showing that populations 1 and 2 form a cluster that is rather different from populations 3 and 4, which form a second cluster.

It may occur to you that you could easily see these patterns by simply looking at the allele frequencies, which clearly showed the relationships between these populations. Why bother with all this excess mathematical detail? For one thing, in real life situations, we deal with more than one gene at a time and thus have tables that may list dozens of allele frequencies for each population. There is simply too much data to comprehend by simple examination of a table of numbers. The computation of genetic distance uses overall averages to get an idea of the average degree of genetic similarity between populations.

Interpreting genetic distances can also get complicated if there are more than a handful of populations in an analysis. Genetic distances are computed between all pairs of populations. In the above example there were four populations and six genetic distances—one between each pair of populations. If we were looking at five populations the number of comparisons would increase to ten genetic distances, and if we were looking at ten populations, we would have forty-five genetic distances. As the number of distances becomes larger, it becomes harder to look at a table of numbers and easily see underlying patterns of relationship.

Here's an example—some hypothetical genetic distances between five populations: A, B, C, D, and E:

Distance between populations A and B = 0.034
Distance between populations A and C = 0.084

Distance between populations A and D = 0.130
Distance between populations A and E = 0.120
Distance between populations B and C = 0.084
Distance between populations B and D = 0.132
Distance between populations B and E = 0.125
Distance between populations C and D = 0.050
Distance between populations C and E = 0.039
Distance between populations D and E = 0.030

These numbers are not as easy to interpret as those in the first example because we are now dealing with more populations and hence more distances. You *can* see some overall pattern if you look at the numbers long enough. Populations D and E are the most similar to each other because they have the smallest genetic distance between them. Likewise, populations A and B have a relatively low genetic distance between them. In addition, population C is more similar to D and E than to A and B. Although this pattern is apparent if you look closely at the individual genetic distances, it may not be immediately obvious with a casual scan of the numbers. If more populations were added to the analysis, then the number of populational comparisons would become too unwieldy. In one analysis I did (discussed in Chapter 9), I looked at the genetic distance between thirty-one different populations, which produced a list of 465 distances. One can go blind (or crazy) trying to make sense out of so many numbers.

The solution is to follow the old adage that a picture is worth a thousand words (or numbers). But how can we draw a picture that represents these ten genetic distances? If we were looking only at two populations, we could represent the distance between them as a line on a one-dimensional graph. If we were looking at only three populations, we could represent the distances between these populations as a two-dimensional plot, with the distance between two of the three populations on one axis, and the distances to the third population on the second axis. Likewise, the genetic distances between four populations could be represented as points in a three-dimensional plot. However, what do we do if we are looking at the genetic distances between five populations? We would need a four-dimensional graph, which we cannot draw or visualize! In general, if we have n populations we need $n - 1$ dimensions to show these distances (this is a maximum; in some cases we would need less). The point is that if we are looking at more than a small number of populations, we need a method that can show a picture of the genetic distances in a smaller number of dimensions.

Fortunately, there are mathematical methods that can reduce a multidimensional plot to fewer dimensions, often without losing much of the original information on population relationships. Such methods are mathematically complex, but they essentially reduce data into a reasonable form that we are capable of drawing and interpreting. We could squeeze the distances down into a one-, two-, or three-dimensional picture. In most cases, I have found that the easiest plot to understand is two-dimensional. Looking at only one dimension usually doesn't show as much information, and I personally find a three-dimensional picture more difficult to interpret.

I refer here to the two-dimensional plots used throughout the remainder of this book as genetic distance "maps," because you read a genetic distance map the same way you would read a road map. When looking at points on a road map, you know that the closer two places are on the map, the closer they are in actual geographic space. The same principle applies here, but the map is based on genetic distance rather than geographic distance. Populations that are genetically more similar are closer together on the genetic distance map, whereas genetically different populations are more distant from each other on the map.

Figure 5.1 is a genetic distance map of the five hypothetical populations that I presented earlier (A, B, C, D, E). The points on the map represent the five populations. Which points are closest together on the map? The closer two points are to each other, the more similar they are genetically, because the map was constructed from the set of genetic distances. One can easily see that populations A and B form a cluster on the plot, indicating that they are more genetically similar to each other than to populations C, D, or E, all of which are farther away on the map. Because of the mathematical method that is used to "squeeze" the multidimensional space into two dimensions, we look first at the horizontal axis, which accounts for more of the variation among genetic distances, and secondarily at the vertical dimension. In this case, the primary distinction is a separation of two clusters along the horizontal axis; one cluster consists of populations A and B, and the other cluster consists of populations C, D, and E. The vertical axis separates populations in the second cluster, showing population C to be significantly different from

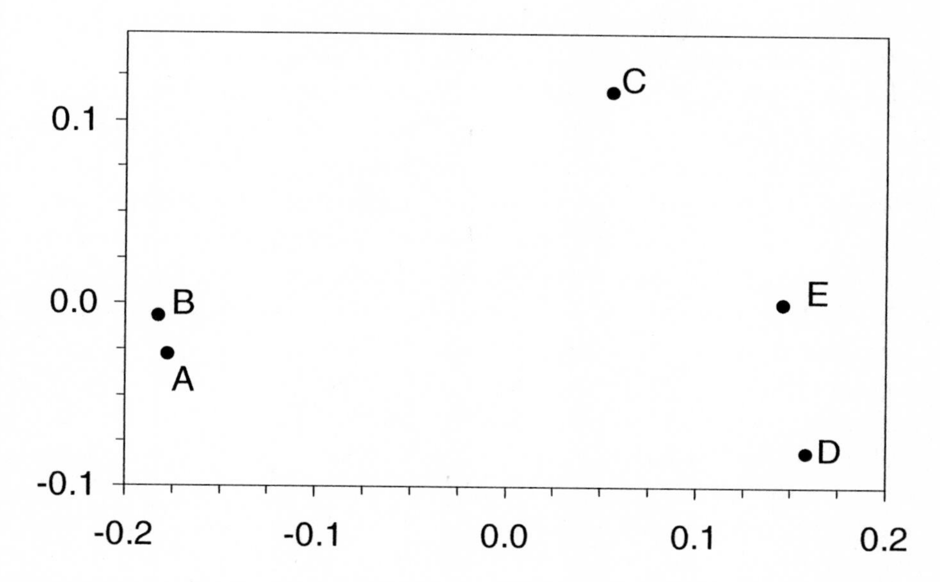

Figure 5.1 Example of a genetic distance map. This two-dimensional plot represents the hypothetical genetic distances between five populations (A, B, C, D, E) given in the text. Genetic distance maps are interpreted in the same way as road maps; the closer two points lie on the map, the closer they are to each other genetically. The numbers along the axes are the coordinates in "genetic space." We can conclude that there are three clusters of genetically related populations; one consists of populations A and B, another consists of populations D and E, and the third contains only population C.

the others. (By the way, it doesn't matter that A and B are on the left side of the plot or that C is at the top; the orientation of genetic distance maps is arbitrary, and what matters is the relative distance between points in the overall two-dimensional space.) We conclude that populations A and B are very similar to each other, as are populations D and E, but these two clusters are distinct from each another. Finally, we also see that population C is different from both of these clusters. I use this type of genetic distance map frequently throughout the rest of this book as a graphic way of summarizing genetic distances.

The final step in the analysis (and often the hardest) is making sense of the results. How can we explain the pattern of genetic distances that we have observed? *Why* are some populations more genetically similar? One possible answer is the impact of geographic distance. Perhaps populations

A and B are located near one another geographically, such that they share more genes with each other than with more distant populations. Perhaps the same thing is true of populations D and E, and population C is geographically distant from all other populations. This can be tested rather easily by comparing the genetic distance map with a map of geographic distances. Do the two maps look similar to each other? On the other hand, the situation might be a bit more complex. Perhaps some populations have a common history that explains their genetic similarity, as would happen if, say, population B was formed by migrants from population A. Genetic differences might also reflect physical barriers. For example, perhaps population C is genetically different because it is isolated by a physical barrier, such as a mountain, which would reduce gene flow. Or, perhaps C is culturally different, and this cultural isolation has reduced gene flow. Another possibility (among many!) is that population C includes a lot of immigrants from another part of the world that is genetically different. These are only some of the possible explanations for the observed genetic distances. How can we choose among these possibilities? One way is to relate these differences to other information on these populations, including geography, history, and culture. The interpretation of genetic distances is a bit like solving a mystery; you need a variety of clues from different sources to solve the puzzle.

Global Genetic Diversity and Isolation by Distance

The focus of the remainder of this chapter is on interpreting genetic distances on a global level, looking for patterns of relationship between human populations across the entire planet. Many such studies have been done in the past few decades, but the most comprehensive analyses to date have been conducted by geneticist Luca Cavalli-Sforza of Stanford University. In 1994, Cavalli-Sforza and his colleagues Paolo Menozzi and Alberto Piazza published a massive volume in excess of 1,000 pages titled *The History and Geography of Human Genes.*[5] Most of this volume deals with the analysis and interpretation of classic genetic marker data collected on hundreds of human populations across the world. Two of their analyses are particularly useful to examine here to review global patterns of genetic distance and set the stage for discussions of regional and local population history in later chapters.

One of their analyses involved looking at the genetic distances among forty-two populations grouped into nine large geographic regions:

1. Sub-Saharan Africa
2. Europe
3. Middle East, North Africa, and South Asia
4. Southeast Asia
5. Australia and New Guinea
6. Pacific Islands
7. Northeast Asia
8. Arctic Northeast Asia
9. Native Americans

In each case, they used data from native populations and excluded groups known to have moved in the recent past, particularly those groups that migrated large distances in the 500 years since European exploration. They then computed the genetic distances between the nine regions using averages based on 120 different alleles. This produced thirty-six pairs of genetic distances, which is too many to interpret meaningfully by just looking at a table of the distances. Instead, I used these distances to produce the simple genetic distance map shown in Figure 5.2.

We see that the horizontal axis of the genetic distance map separates the Australian and New Guinean population at one end and the sub-Saharan African population at the other, with Asian and European populations plotting in between. The vertical axis separates Native American and Asian populations. Overall, we can see several patterns on the genetic distance map. Sub-Saharan Africans are the most genetically distinct. Native Americans are most similar to Arctic Northeast Asians. Australians, New Guineans, and Pacific Islanders are most similar to Asians, particularly Southeast Asians. Although we are losing a lot of potentially valuable information by lumping many individual populations into regional aggregates, there are still patterns that might reflect global population history to some extent. But what exactly do these patterns mean?

One of the first things we need to investigate is the relationship of the genetic distance map to geography. Genetic similarity is affected by levels of gene flow; the greater the level of gene flow between two populations, the more similar they are to each other. A basic rule of thumb in most population genetic studies is that gene flow is affected by geographic distance.

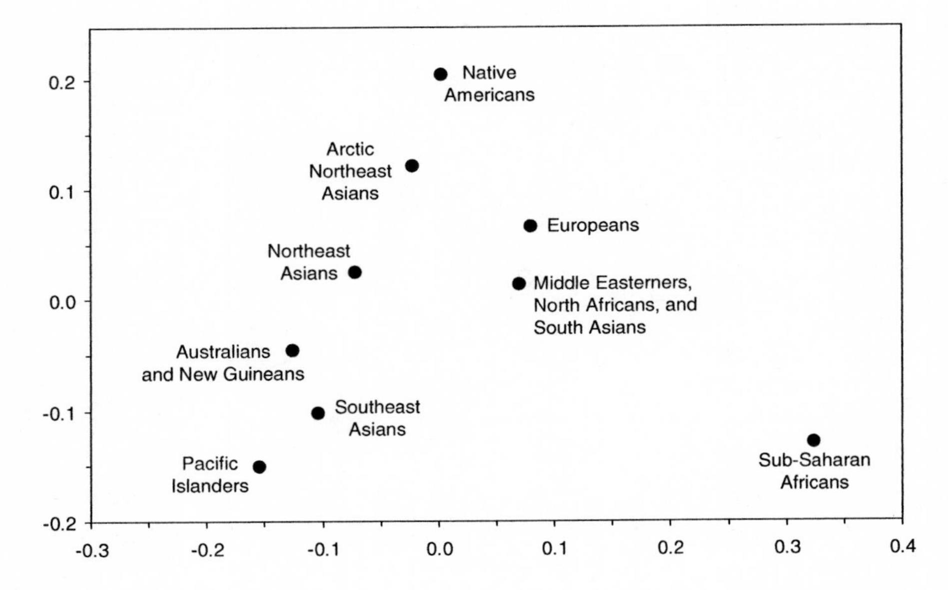

Figure 5.2 Genetic variation among human populations. This graph shows the genetic distance map for nine groups of living humans based on 120 alleles. This map provides a two-dimensional picture of the genetic relationships of human populations across the world. Genetically similar populations plot close to each other, such as Southeast Asians and Pacific Islanders.
Source: Cavalli-Sforza et al. (1994).

Quite simply, populations are more likely to share genes with those geographically close by than with those farther away. We therefore expect, everything else being equal, that there will be some correspondence between genetic distance and geographic distance. Although this is not always the case, it is a good place to start. We can see some influence of geographic distance in Figure 5.2. For example, the positions of the Asian populations on the genetic distance map clearly correspond, north to south, with geography. The aggregate group that includes Middle Easterners, South Asians, and North Africans lies between sub-Saharan Africans and East Asians, which is also expected based on geography. Another example of correspondence between genetics and geography is the similarity of Australians/New Guineans and Pacific Islanders to their closest neighbors in Southeast Asia.

We need to consider history as well as geography. Geography has structured the migration of groups into different parts of the world and influenced the historical relationships between populations. We know that location and terrain affected patterns of human movement in the past. Archaeological and genetic evidence shows that the first Americans came from Northeast Asia (discussed in Chapter 6), most likely moving across the Bering Strait that separates the Asian and North American continents. Likewise, we have evidence that humans moved into Australia and later into the Pacific Islands from Southeast Asia. Humans moving out of (or into) Africa would have most often exited and entered from the northeastern part of the continent across the Suez Isthmus. If we put all of this together, we begin to get a picture of the actual geographic distances involved in both colonization and genetic exchange. For example, the geographic route between sub-Saharan Africa and the Americas would not simply be the straight line distance across the Atlantic Ocean between these two points but would be constrained along a path that led northeast out of Africa, across Eurasia to Northeast Asia, and then across the Bering Strait and south into the Americas.

If we consider all of these factors, we get a picture that looks like the schematic presented in Figure 5.3, which has the Eurasian landmass connecting to three different regions: sub-Saharan Africa, Australia and the Pacific Islands, and the Americas. In other words, the gene flow from Africa to the Americas would have occurred not across the relatively short distance of the Atlantic Ocean but across the vast Eurasia landmass. The same thing applies when connecting any of the three "legs" in Figure 5.3. The genetic distance map shows a strong correspondence with this picture. Eurasian populations are connected to three regions: Africa, Australia and the Pacific Islands, and the Americas. This is even more apparent in Figure 5.4, in which the Eurasian groups and the three "legs" have been shaded on the genetic distance map.

From these analyses it is clear that human genetic distances show some correspondence with geography, particularly when we factor in what is known about possible routes for colonization and migration. In other words, the genetic distance map gives us a picture of global human diversity that has been shaped by both geography and history. However, while useful, these analyses are somewhat crude because they lump populations into rather large geographic regions. They show a general pattern, to be sure, but they mask relationships between the smaller local populations.

Cavalli-Sforza and his colleagues created a genetic distance map that reveals the finer details of population relationships. Figure 5.5 shows the

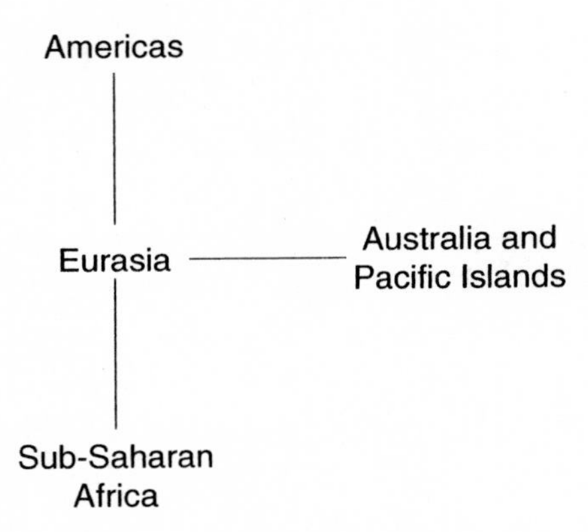

Figure 5.3 Geographic distances between regions of the world. This is a schematic representation (not to scale) illustrating likely paths of movement across the world in earlier times. Genetic exchange would frequently involve travel across Eurasia. For example, the likely path of population movement, and thus gene flow, between Africa and the Americas would not have been a straight-line connection across the Atlantic Ocean, but a path across the Eurasian landmass.

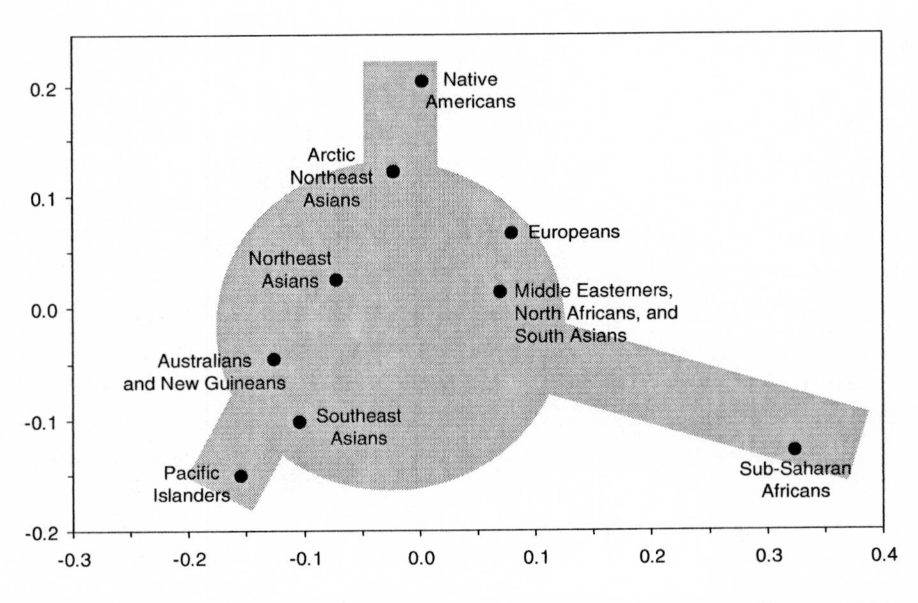

Figure 5.4 Global correspondence between geography and genetic distance. This genetic distance map is the same as the one shown in Figure 5.2, but with the addition of shaded areas showing correspondence to the schematic of geographic distance in Figure 5.3.

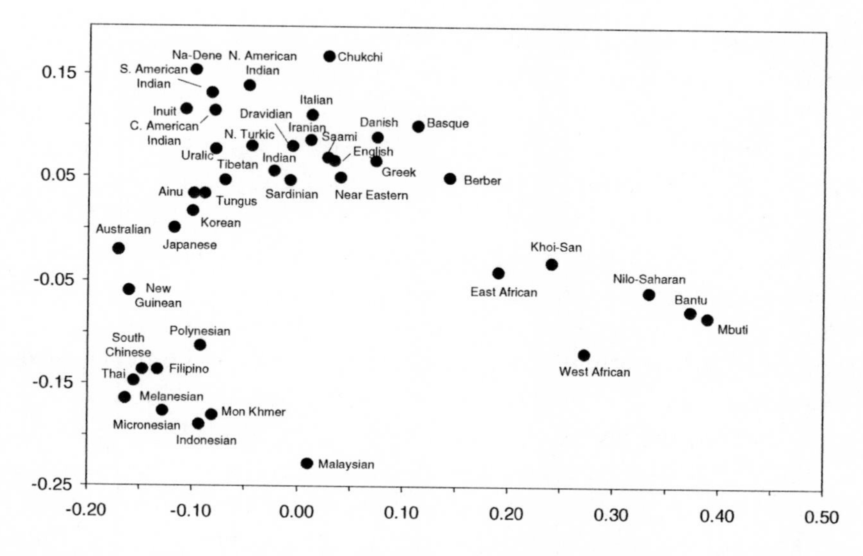

Figure 5.5 Genetic variation among human populations. This graph shows the genetic distance map for 42 local populations of living humans based on 120 alleles. This map provides a two-dimensional picture of the genetic relationships of human populations across the world. The overall picture is the same as shown in Figure 5.2 but with greater detail on variation within geographic regions.
Source: Cavalli-Sforza et al. (1994).

genetic distances for forty-two populations chosen to represent different geographic, linguistic, and cultural groups within humankind. The basic geographic and historical patterns seen in the earlier analysis are apparent in this map as well; Eurasian populations tend to fall in the upper third of the map, with the Australian/New Guinean, Pacific Islander, Native American, and sub-Saharan African populations surrounding this core. Explanations for the relative placement of the local populations are not as quickly apparent. Why, for example, are Polynesians closer on the genetic distance map to the South Chinese than to Melanesians, who are actually closer geographically? Much of the remainder of this book deals with the relationship between genetics and history within geographic regions. The genetic distance map in Figure 5.5 provides only a broad picture of these relationships.

These analyses suggest that geographic distance has been a major determinant of genetic distances between populations. This is not unexpected; population geneticists have noted that genetic diversity in many species follows what is known as the isolation by distance model.[6] According to this model, genetic distances are expected to increase with increasing geographic distance between populations. This increase is rapid at first, and then it slows down, reaching a plateau where there are few changes in genetic distance (see Figure 5.6).

To see how well this model fits the genetic distances obtained by Cavalli-Sforza and his colleagues, I computed the geographic distance between each pair of the forty-two populations in their analysis. To provide a more reasonable estimate of the actual distances separating populations, I made a number of adjustments. All distances between Native American populations and those in other regions reflect a crossing of the Bering Strait, since that was the point of entry into the Americas. Likewise, I adjusted the distances for Australia and the Pacific Islands so that they reflect a path through Southeast Asia, and distances between sub-Saharan Africa populations and other geographic populations reflect a path through northeastern Africa. This method is not perfect, but it provides a more accurate first approximation to actual geographic distance than simply measuring distance by a straight line connecting two geographic points.

I then plotted genetic distance against geographic distance for each pair of the forty-two populations—a total of 861 comparisons. Figure 5.7 shows a plot of each of these 861 points; each point represents the geographic and genetic distance for a pair of populations (for example, Malaysia and New Guinea). There is a noticeable increase in genetic distance with increasing geographic distance, although there is also considerable scattering in the plot—the correlation of the data to the model is not perfect. This suggests that isolation by distance has had an important effect on human genetic distances, but that it isn't the only factor.

Another way of assessing the isolation by distance model is to look at the *average* genetic distances within different classes of geographic distance. Here, I took all the genetic distances between pairs of populations that were geographically separated by a distance of between 0 and 499 kilometers and calculated an average for this distance class. I then did the same thing for genetic distances corresponding to populations geographically separated by distances of between 500 and 999 kilometers, and so forth, to arrive at forty-eight total intervals of geographic distance (based on 500-kilometer increments). This provides a picture of the *average* relationship between genetic and geographic distance, shown in Figure 5.8, which shows a strong correlation to the isolation by distance model. We

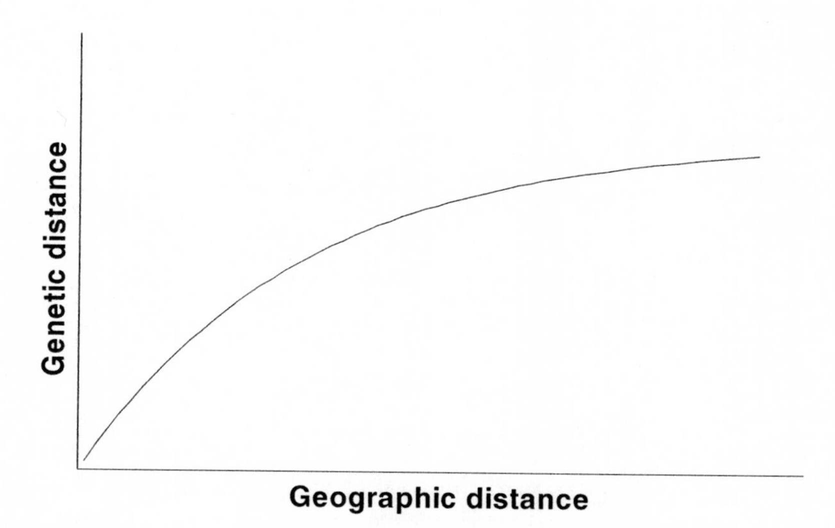

Figure 5.6 Isolation by distance. The theoretical model predicts a relationship between the geographic distance and genetic distance between populations. Populations that are located farther from each other geographically are expected to be genetically more dissimilar. Genetic distance increases with geographic distance. This increase tends to level off at very large geographic distances.

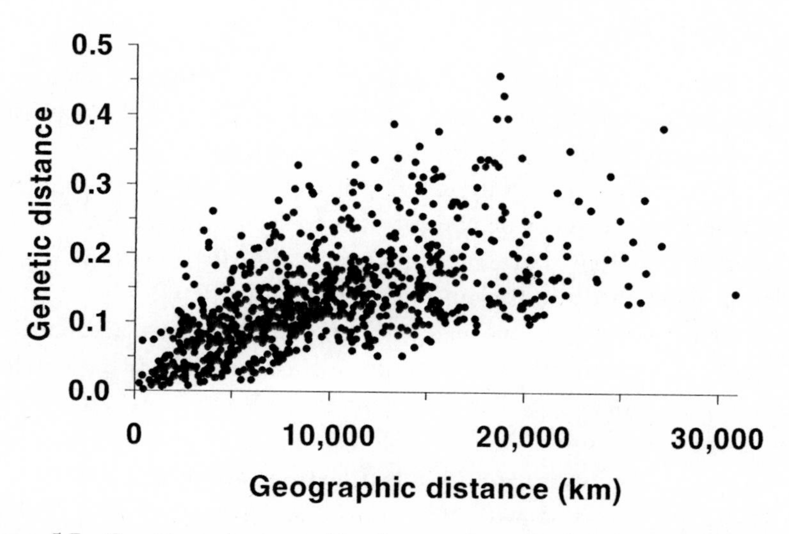

Figure 5.7 Genetic and geographic distances between human populations. Each point represents the geographic distance (horizontal axis) and genetic distance (vertical axis) between a pair of human populations. The entire plot shows all of the pairings for 42 human populations across the world. These are the same genetic distances used to derive the genetic distance map in Figure 5.5.

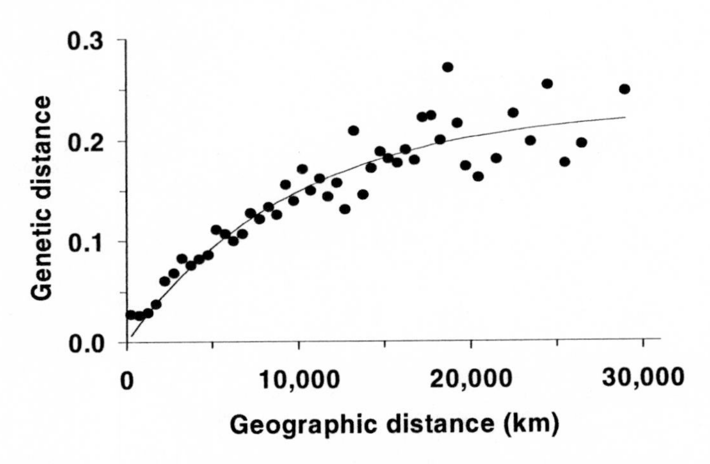

Figure 5.8 Genetic and geographic distances between human populations. This plot was made by averaging the values shown in Figure 5.7 into different classes based on the geographic distance between populations. This plot shows the *average* relationship between genetic and geographic distance more clearly than in Figure 5.7. For geographic distance less than 20,000 kilometers, average values were grouped by 500-kilometer intervals (0–499 km, 500–999 km, and so on). For geographic distances between 20,000 and 27,000 kilometers, average values were grouped by 1,000-kilometer intervals to increase sample size (20,000–20,999 km, 21,000–21,999 km, and so forth). The few distances greater than 27,000 kilometers were all pooled.

can therefore conclude that the geographic distance between human populations, constrained by the history of migration, has had a major influence on the patterns of genetic distance in our species.

In short, populations that are geographically closer to each other, such as Greece and Italy, tend to be more similar to each other genetically than populations that are far apart, such as Greece and Micronesia. Again, this is common sense, because the amount of gene flow between populations affects their similarity to each other, and geographic distance restricts gene flow. This was certainly true throughout most of human evolution, where migration was limited to walking. Even in today's world, where we have greater mobility because of our technology, we tend to choose mates from the local population; these are the individuals we are most likely to meet on a frequent basis and with whom we establish close relationships. Opportunities for interacting with, or mating with, people from distant populations are much more limited.

On the other hand, there has been (and still is) gene flow across the entire planet. Some of this gene flow results from movement from one local population to the next, generation after generation. A few migrants might move into the next village, which in turn shares some migrants with the next village down the line, until, after many generations, genes have flowed across long distances. There are also individuals who move great distances, perhaps because of colonization, invasion, or other forms of long-distance movement. There have been episodes of both short-range and long-range movement throughout human history and prehistory.[7] One of the goals of anthropologists interested in genetic history is to figure out the differential impact of short-range migration and long-distance dispersal and mixture. A number of examples of such studies are described in later chapters.

Genetic Diversity Between Populations and Individuals

The flow of genes across long distances becomes important when considering another facet of human genetic diversity: We are a fairly homogeneous species. Although some aspects of our physical features appear quite variable, such as skin color, most of what we see genetically is a narrow range of diversity in our species. For some traits, there are fewer genetic differences among humans across the globe than there are among neighboring populations of chimpanzees.[8] For most traits, we simply are not that different from one another relative to other primate species.

One useful means of looking at the heterogeneity of our species is to examine how much of our total genetic diversity exists at different populational levels. One way of doing this is to divide the species into a number of large geographic groups, such as sub-Saharan Africans, East Asians, and so forth, and to determine how much of our total genetic diversity exists *between* regional populations and how much exists *within* regional populations.

To understand how this works, consider a simple example. Imagine a species made up of two regional populations. Everyone in the first regional population is a clone of a given person; that is, everyone is genetically identical. In this case, there is no genetic diversity within the population because everyone is the same. Now, imagine that the same thing applies to the second regional population; that is, everyone is a clone of one individual (but a different individual than the one in the first regional population). Again, there is no genetic diversity within the region. However, because the two regional populations are made up of clones of genetically different people, *all* of the diversity in this species exists *between* regions (100 percent) with no diversity *within* regions (0 percent).

Now, consider a completely different situation. Each regional population is composed of genetically different people, but the proportions of different traits is the same in both populations. For example, consider a case where, in each population, 85 percent of the people have the Rhesus positive blood type and 15 percent have the Rhesus negative blood type. Because some people are Rhesus positive and some are Rhesus negative, there is variation *within* both populations. However, there is no genetic difference *between* the regional populations because the relative proportions are the same in both. In this case, none of the genetic diversity exists *between* regions (0 percent), and all of the genetic diversity exists *within* regions (100 percent).

Species that consist of geographically isolated populations tend to exhibit greater variation *between* regional populations than species that have a history of migrating populations. Where do humans fit in? Does most of our genetic diversity exist between regional populations or within regional populations? A number of studies have demonstrated that, for most human genetic traits (and some physical ones), only about 10 percent of our diversity exists between regional populations, with the remaining 90 percent existing within regional populations.[9]

The partitioning of diversity can also be extended to consider sources of variation within regions. Each regional population, such as sub-Saharan Africans, is made up of local populations, such as the Mbuti or the San, and each local population is made up of individual humans. How much genetic difference exists between the local populations within regional populations, and how much exists between the individuals within local populations? Studies have found that, on average, about 5 percent of our genetic diversity exists between local populations within regional populations, and 85 percent exists between individuals within local populations (Figure 5.9).[10]

Most of the genetic differences between people in the world exist between individuals within populations, with much less overall difference between populations. This conclusion runs counter to the common use of select physical traits, such as skin color, to separate humanity into different races. Skin color *does* show more diversity between populations than within them; the skin color of any two natives in a rural village in Scotland will be more similar than the skin color of the rural Scot and that of

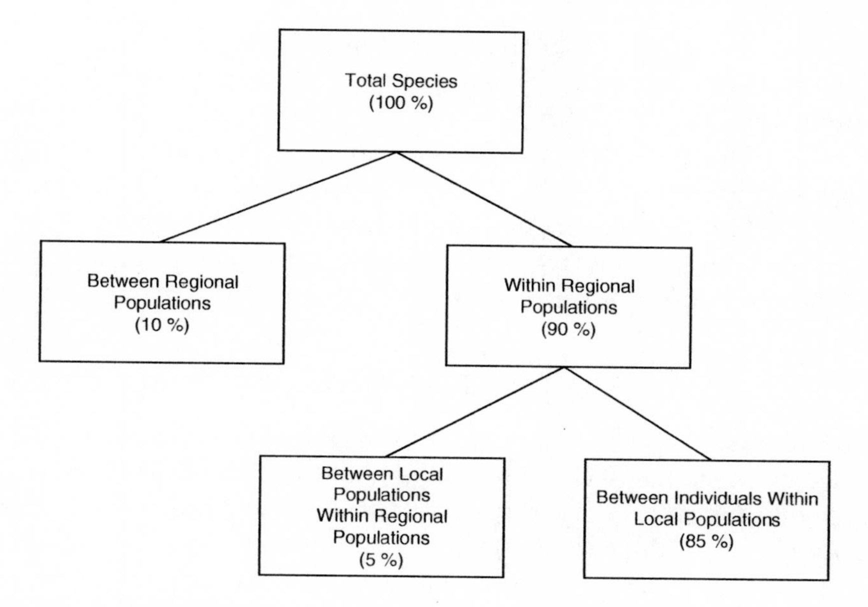

Figure 5.9 Apportionment of human genetic diversity. This schematic represents the partitioning of genetic diversity in the human species based on analyses of classic genetic markers, DNA markers, and cranial and facial measurements. The overwhelming amount of human genetic diversity exists between individuals within local populations.

a native of a village in Kenya. We tend to divide humanity into different groups (races) based on such features, since it is a trivial matter to classify native Scots and Kenyans into two groups based on their skin color (although if we were to add more groups to this analysis it would get harder to draw distinct boundaries between human populations, because human skin color varies *continuously* across space).

The problem comes when extrapolating from one trait (skin color) to the pattern of diversity in our species for all other traits. This gets particularly problematic when one attempts to extend such generalizations to behavioral traits, such as IQ test scores, which are clearly influenced by environmental factors that vary across populations. Studies of the partitioning of human diversity show that skin color is atypical, and results from blood groups, protein and enzyme markers, DNA markers, and even cranial measurements show a completely different pattern.[11] These traits are capturing information on the genetic structure of our species in terms of our population history, whereas skin color is telling us only the history of that particular trait. Skin color shows greater differences across geographic regions because of natural selection and adaptation to different latitudes; skin color is darkest at or near the equator and becomes lighter with increasing distance away from the equator.[12] On a global level, skin color tells us little about the history of populations; instead it tells us about the history of natural selection for skin color. It is ironic that skin color has been used so frequently in attempts at racial classification and population history.

The Palimpsest Revisited

We can see some basic patterns when looking at human genetic diversity at a global level. Eurasian populations are centrally located on the genetic distance maps, which reflects that fact that Eurasia is a geographic center connecting more distant regions. The population of sub-Saharan Africa is most similar to geographically proximate populations in the Middle East and North Africa. The populations of Australia, New Guinea, and the Pacific Islands are closest genetically to the population of Southeast Asia, which is also the closest geographically. Most of the patterns we see on a global level relate to geographic distance. The farther apart human populations are, the less gene flow between them, and the more genetically different they become.

The patterns of genetic distance between human populations have also been shaped by the specific history of colonization and expansion, which in some cases has been geographically constrained. We can conclude, in a general sense, that Pacific Islanders came from Southeast Asia and that Native Americans came from Northeast Asia. At this global level, we are just beginning to peel back the layers of the palimpsest of human diversity. A more detailed examination will reveal the text that lies beneath this general picture.

SIX

The First Americans

Most of us are familiar with at least part of a poem learned in grade school that begins "In 1492, Columbus sailed the ocean blue." This short line helps us remember the story of the voyages of Christopher Columbus and the period of European exploration that resulted in contact between the Old World (Africa, Asia, Europe) and the New World (the Americas). The story I was taught as a child stressed the scientific importance of Columbus's first voyage, namely, that Columbus was convinced that the earth was round and set off to prove it by circumnavigating the globe. Actually, this story is a bit of a myth, as virtually everyone in Columbus's time (and earlier) accepted that the earth was round. In fact, one of the objections raised by a commission formed by King Ferdinand and Queen Isabella to review Columbus's plans was that the earth's circumference was too great for him to reach his intended goal (the East Indies) in the length of time he proposed.[1]

As we all know, Columbus sailed westward from Spain on August 3, 1492, and came to land in the Americas on October 12, 1492. He was convinced that he had indeed gone around the world and landed somewhere in Asia, a region known in Spain at that time as the "Indies," thus prompting him to name the native people "Indians." It turns out that his critics in Spain were correct; the earth's circumference was greater than Columbus had estimated. Rather than landing in Asia, he had landed on a small island in the Bahamas. After this and subsequent voyages by Columbus and others, the world changed dramatically as contact increased between the Old and New Worlds, whose human populations had hitherto been separated for many thousands of years.

Was this the first contact between the Old World and the New World? There have been many suggestions that contact took place before Columbus, including supposed voyages from Ireland, Africa, and China, but these have been rejected due to a lack of archaeological evidence. However, there is evidence to support a Viking exploration of North America some 500 years before Columbus's voyages, although contact appears to have been quite limited, and the Vikings never established a large colony.[2] In terms of enduring historical significance, however, the many social, political, and economic changes resulting from culture contact began with Columbus's voyage in 1492.

Almost immediately, Europeans began to wonder about the origins of these newly "discovered" peoples. Where did they come from, and when did they get there? Today, evidence from biology and archaeology has answered this question. The first Americans were migrants dispersing out of Asia about 15,000 years ago (and perhaps earlier). Although we have the general answer, there are still many unanswered specific questions about the origins of Native Americans. The three major questions consist of the following:

1. Where in Asia did Native Americans come from?
2. How many migrations took place?
3. When did all of this happen?

This chapter examines the role that genetic data has had in answering these questions about the population history of the first Americans.

Where Did the First Americans Come From?

When Europeans colonized the Americas, they found a wide range of people and cultures in North, Central, and South America. Some native groups were small and subsisted as hunters and gatherers. Others lived in larger agricultural societies and some in cities. Archaeologists would eventually provide evidence of prehistoric civilizations and empires, including the Maya, the Aztecs, and the Inca, among others. There was clearly a long history of occupation of the Americas that was, until the beginning of European exploration, unknown to the Western world. Where did these natives come from?

In the first few centuries following Columbus's voyages, most European scholars relied on biblical interpretation to explain the origin of Native Americans, reasoning that these natives were the descendants of the lost tribes of Israel alluded to in the second Book of Kings in the Bible.[3] This interpretation provided a solution to a sticky problem for those taking the Bible as a literal and accurate historical record. By Columbus's time, biblical scholars had viewed racial diversity in humans as having originated with the dispersion of the three sons of Noah (Japheth, Ham, and Shem) following the flood described in the Book of Genesis. According to this view, Japheth's descendants gave rise to Europeans, Ham's descendants to Africans, and Shem's descendants to Asians. Since Noah had only three sons, then the newly discovered race of Americans must have descended from one of these three. But which one? The biblical mention of the ten lost tribes of Israel (presumably themselves descendants of Shem) was an answer that was consistent with a literal interpretation of the Bible. Although the biblical interpretation was popular for some time, other suggestions were also made, including transatlantic voyages from Egypt and even an origin in mythical Atlantis.

Not everyone attempted to answer the question of American origins using biblical scholarship. Friar Joseph de Acosta, writing in 1590, took a different tack and approached the problem by looking at the geographic distribution of different animal species. He noted that there was evidence that many animal species had migrated from the Old World to the New World at different times in the past. If this were true of other species, de Costa reasoned, it could also be true of humans. He inferred that there must have been a geographic connection between the Old and New Worlds to allow this migration and, based on his geographic knowledge, the most likely place was in the northeastern part of Asia and the northwestern part of North America. This location is important because of the close proximity of the two continents separated only by the Bering Strait (Figure 6.1). At no other point are the Old and New Worlds geographically closer to each other.

Although for a long time most people favored the biblical interpretation, eventually scientific data would confirm de Acosta's idea. There is an abundance of biological data from both prehistoric and contemporary populations showing that the first Americans came from Asia. These data show that sometime after humans expanded into Northeast Asia, some group(s) continued to expand eastward into the Americas across the Bering Strait. The traditional explanation has been that hunting and gathering populations moved directly into North America during the last ice age when the New and Old Worlds were not separated by water but connected by the Bering land bridge.

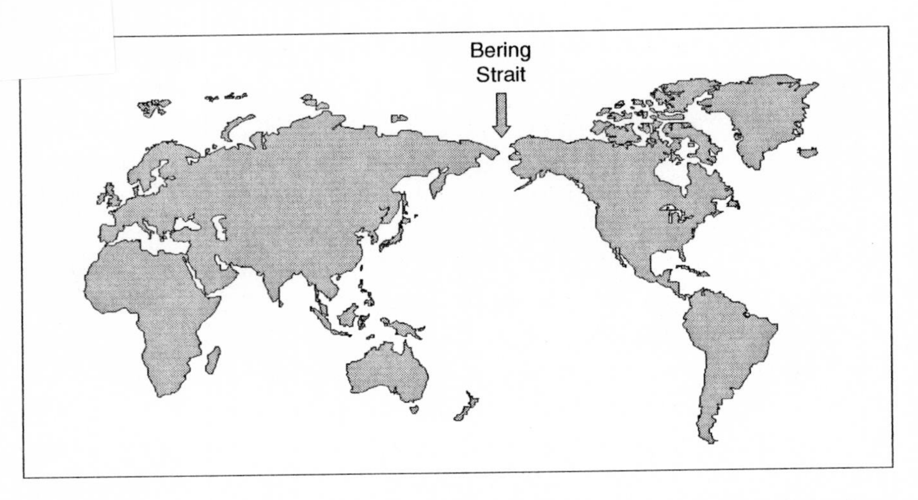

Figure 6.1 Location of the Bering Strait, showing the close proximity of Northeast Asia and North America. Traditional archaeological explanations for the origin of the first Americans propose that humans from northeastern Asia crossed over the Bering land bridge—a physical connection between Asia and North America that was exposed when the sea level dropped during an ice age—and then expanded into the New World. Recent debate suggests the possibility that migrants may have used boats to reach the New World and then traveled south into the Americas along the western coast of North America.

What does an ice age have to do with migration from the Old World to the New World? Water normally evaporates into the atmosphere and then returns to the oceans through precipitation. Much of the water is frozen during an ice age and cannot flow back into the oceans; although evaporation continues to take place, the water does not return to the oceans but remains on land as snow and ice. This trapping of water causes sea levels to drop, which in turn uncovers connecting land between continental landmasses that previously were under water. Humans moved across the land bridge into North America. Once there, there were two major glacial ice sheets preventing further movement to the southwest and east. However, there was an ice-free corridor that was open at certain points in the past, and it has long been thought that humans moved through this corridor down into North America and a rich land previously not seen by human eyes. Humans then spread rapidly throughout the New World, occupying land from the northernmost parts of North America to the tip of South America.[4]

The term "bridge" may not convey the most accurate image of what occurred. One should not imagine a group of people rushing single file across a narrow strip of land connecting Asia and North America! In reality, the "bridge" was almost 1,300 miles wide. And this was not a sudden event, where the waters receded like a bathtub full of water when the plug is pulled, followed by a mad rush across the mud to the New World! The geologic changes brought about by an ice age occur over many, many years, and the habitat on the land bridge changed very slowly. Eventually forests expanded across this stretch, as did herds of animals. It is unlikely that the first migrants to the New World knew they were crossing a "bridge," but rather they expanded naturally into new territory, perhaps following game herds. When anthropologists speak of a migration, they are suggesting not a sudden movement of people but rather an avenue for dispersal that may have been used for many generations.

This traditional view of movement across the Bering land bridge and subsequent entry into North America via the ice-free corridor has been challenged in recent years. Several archaeologists have proposed that humans entered the New World through the use of boats and that much of the early migration was movement south along the western coasts of the Americas. There is no reason that this couldn't have happened; humans made and used boats at least 60,000 years ago to reach Australia, a continent that was not connected by a land bridge to other continents. It is also possible that both ideas are correct, that some humans moved across the land bridge on foot and others spread by water. In any case, what is clear is that these humans came primarily from Asia, a conclusion firmly in line with the genetic evidence.

To put this into evolutionary context, keep in mind that the first Americans were anatomically modern humans. Many archaeological investigations of New World sites have yielded skeletal remains. *All* of these skeletal remains are anatomically modern humans. There are no Neandertals or other archaic humans, no remains of *Homo erectus*, and no remains of early hominids. Neandertals and other archaic humans had lived thousands of years before humans entered the New World. All previous stages of human evolution took place in the Old World, and by the time humans entered the New World, they were already fully modern anatomically.

The Genetic Link Between Asia and North America

Patterns of human biological diversity support de Acosta's idea of an Asian origin of the first Americans. For example, there are several average physical similarities between Native Americans and East Asians, including straight black hair, relative lack of facial hair, broad cheekbones, and a higher incidence of the epicanthic fold, the fold of skin across the inner part of the eye. These traits are not confined to these groups, nor does everyone in these groups share these traits. A number of exceptions show that it is not easy to separate humanity into different groups based on the presence or absence of a trait. For example, some African populations show an epicanthic fold, and some East Asian populations show a high incidence of heavy beards. Overall, however, the average similarities in certain physical traits suggest a historical connection between East and Northeast Asia and the Americas.[5]

A good example of a physical trait that supports this historical connection is the frequency of shovel-shaped incisors in different human populations. Incisors are the flat front teeth in a mammal; humans typically have four incisors in both their upper and lower jaws, and the incisors in the upper jaw are larger. Many incisor teeth are relatively flat when viewed from the chewing surface, but some people have ridges on the outer margins of the tooth, a condition known as a shovel-shaped incisor. Although shovel-shaped incisors are found to some extent across the world, the *relative frequency* is considerably higher in both East Asian and Native American populations. For example, the frequency of shovel-shaped median incisors (the ones in the middle of your jaw) is about 8 percent in American whites and 12 percent in American blacks but is over 90 percent in East Asian and Native American populations.[6]

The Asian connection is also apparent when looking at classic genetic markers. One example of a genetic marker that shows strong similarity between East Asian and Native American populations is the Diego blood group. As noted in the previous chapter, there are many different blood group systems in the human species. The Diego blood group gene has two different forms (alleles): DI*A and DI*B. (Note that since everyone has two alleles, one from each parent, some people have two DI*A alleles, some have two DI*B alleles, and some have one of each.) The DI*A allele is absent in African and Pacific Island populations, and it tends to be very rare in Europe and the Middle East. Higher frequencies are generally found in both East and Northeast Asian *and* Native American populations (Figure 6.2). Although there are local populations in these regions that lack the allele, the higher *average* frequencies of the DI*A allele in these regions suggest a historical link between East and North Asian and Native American populations.

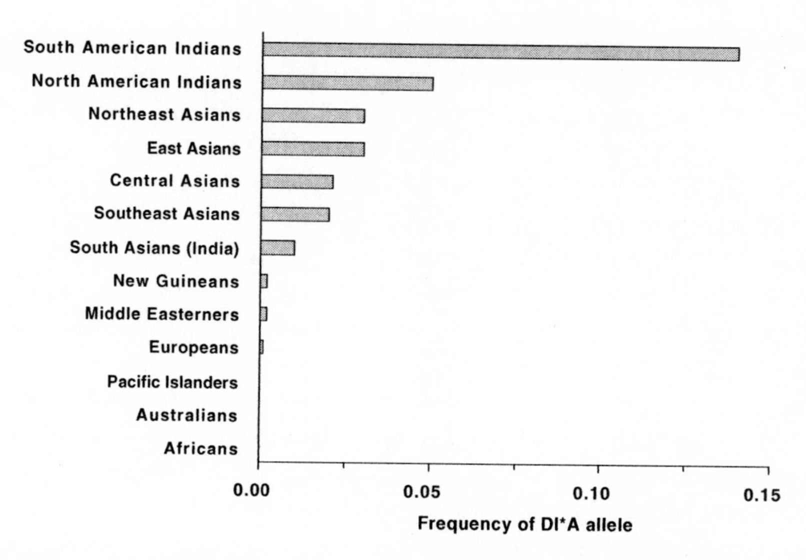

Figure 6.2 Average frequency of the DI*A allele of the Diego blood group in different parts of the world. Populations in the New World, Northeast Asia, and East Asia tend to have higher frequencies of DI*A than those found in other geographic regions, consistent with an Asian origin of Native Americans.
Source: Cavalli-Sforza et al. (1994).

Although these examples are certainly suggestive, are they sufficient by themselves to demonstrate a historical link between Asians and Native Americans? Perhaps shovel-shaped incisors and the Diego blood group are the exceptions rather than the rule. As I noted in previous chapters, not all biological traits provide the same picture. Natural selection can make distantly related populations appear more similar (as in the previous examples of skin color) and make closely related populations appear less similar. There is also the ever-present problem of sampling error; we should avoid making too much out of any single trait, but instead focus on the more statistically accurate picture that can be achieved by averaging results. In the case of the origin of the first Americans, we should examine *average* patterns of genetic distance, based on many different traits. This has been done in a variety of analyses, ranging from studies of cranial shape to blood groups to DNA markers, and the results consistently support a close genetic and historical link between the Americas and Asia, specifically northeastern Asia.

One good example of this type of genetic distance study, described in the previous chapter, is the comprehensive analysis performed by Cavalli-Sforza and his colleagues using frequencies of classic genetic markers from populations across the world.[7] These results show a close genetic relationship between Native Americans and Asian populations, particularly Arctic Northeast Asians (see Figure 5.2, for example). This can be seen more clearly by looking only at the genetic distances of other groups of humanity to Native Americans. Figure 6.3, from Cavalli-Sforza's analysis, shows these genetic distances. The shortest genetic distance (showing the closest similarity) is to Arctic Northeast Asians, followed by other Northeast Asians. As with other traits and analyses, the connection between Northeast Asia and the Americas is quite clear.

Although studies of classic genetic markers confirm the broad picture of a Northeast Asian connection, by themselves they do not provide great specificity. Additional insight has been gained in recent years through the analysis of mitochondrial DNA. By focusing on mitochondrial DNA, with its strict maternal inheritance, we avoid the problems of traits inherited from both parents, where there is recombination of the genetic material each generation. By looking at the evolutionary history of mitochondrial DNA, we can get a clearer picture, at least through the maternal line, of the origins of the first Americans.

One of the ways in which mitochondrial DNA is used to reconstruct population history is through the geographic analysis of mitochondrial *haplotypes*, which are combinations of genes that are inherited together as a single unit (the root "haplo" means single). All of your mitochondrial DNA is inherited as a single unit from your mother. Haplotype analysis starts by using the restriction fragment length polymorphisms (RFLPs), which were described briefly in the previous chapter. DNA is exposed to certain bacterial enzymes that identify specific configurations in the DNA sequence known as restriction sites. There are a number of different restriction enzymes, each one targeted to a specific sequence of DNA bases. For example, the restriction enzyme *Eco*RI is targeted to the 6-base-pair sequence GAATTC and cleaves this sequence between the G and the first A. In other words, if the DNA sequence you are analyzing has the sequence GAATTC somewhere in it, the *Eco*RI restriction enzyme will find it and cut the sequence into two pieces, one containing the base G and the other containing the sequence AATTC. Suppose, however, that there has been a mutation in a given DNA sequence that you are analyzing and the sequence GAATTC has been changed to, say, CAATTC. The restriction enzyme will not find the target sequence, and as a result, the

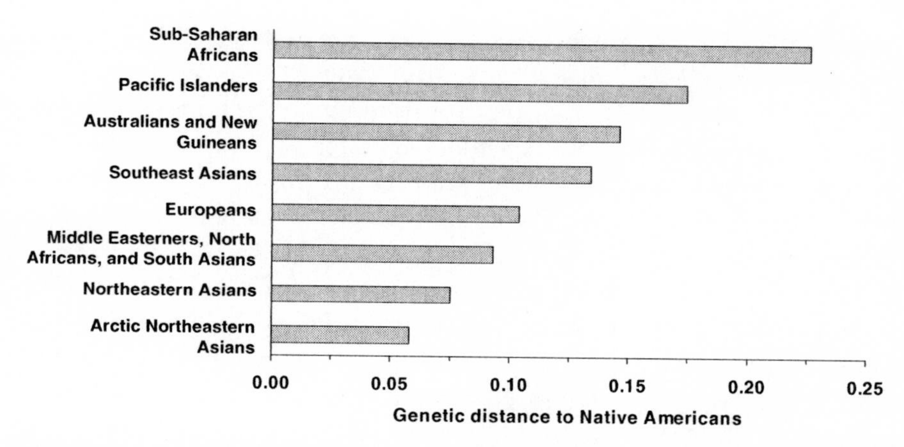

Figure 6.3 Genetic similarity to Native Americans. The genetic distance of Native Americans to other regional populations of the world is shown averaged over 120 alleles. Native American populations are most similar genetically to Northeast Asian populations, particularly those in the Arctic, which reflects the Northeast Asian origin of the first Americans.
Source: Cavalli-Sforza et al. (1994).

DNA sequence will *not* be cut. For any given specimen, we observe whether the restriction site was present (+) or absent (-).

Haplotype analysis is designed to find different combinations of restriction sites. For example, suppose we are looking at three different restriction sites (1, 2, 3), each targeted with a different restriction enzyme. For a given specimen, we might find a match for restriction site 1 (+), a match for restriction site 2 (+), but no match for restriction site 3 (-). We would refer to this specimen's haplotype as + + -. Likewise, a DNA sequence that has a match for restriction site 1 (+) but none for either site 2 (-) or site 3 (-) would have the haplotype + - -. There are, of course, other possible haplotypes in this example, such as + + +, - + +, and so on. Using sophisticated methods, geneticists then look at the genetic relationship between all of the haplotypes to find combinations that share similar mutations. In other words, they build a family tree linking the different haplotypes and then look for clusters of related haplotypes, called *haplogroups*, referred to in the scientific literature by different code letters, such as A, H, X, and so forth.[8]

A rough analogy that helps explain the difference between haplotypes and haplogroups is the make and model of automobiles. There are a number of automobile manufacturers, such as Ford, General Motors, and Honda. Each of these makes includes a number of different models;

for example, Honda models include the Civic and the Accord. Here, automobile models are analogous to haplotypes. The various models of automobiles produced by a given manufacturer are to some extent similar, as they often have similar components, such as the climate controls, basic instrument panel, and other features. In this sense, the different models are related (because of common design), and the different makes of automobiles are analogous to haplogroups.

Mitochondrial DNA haplogroups show an interesting geographic distribution across the world. Some haplogroups are found exclusively or predominately in certain geographic populations. For example, haplogroups L1, L2, and L3 are found in sub-Saharan Africans, whereas the seven haplogroups U, X, H, V, T, K, and J are generally found in Europeans.

Most of the mitochondrial DNA of living Native Americans belongs to one of four haplogroups—A, B, C, and D.[9] These same haplogroups have also been found in ancient DNA extracted from skeletal remains of prehistoric Native Americans. In addition, these haplogroups are also found in Northeastern Asian populations but are rare or absent elsewhere in the world. This shared uniqueness is one of the most striking proofs of an Asian origin of Native Americans. Closer examination of the underlying DNA sequences of these haplogroups shows that Northeastern Asians and Native Americans share certain mutations, a finding best explained by these mutations having occurred in Northeast Asia prior to the migration of the first Americans. As humans moved into the New World, either across the Bering land bridge and/or along the western coast of North America, they carried these mutations with them.

Do these results necessarily mean that Native Americans are *exclusively* of Asian origin? Although most Native American mtDNA haplotypes belong to one of the four (A–D) haplogroups, not all do. Some Native American mitochondrial DNA has haplogroups more typical of Europeans, such as H and J. Many of these non-Indian haplotypes are probably the result of some mating with Europeans after 1492. One exception is haplogroup X, which is found in a number of North American Indian populations and in parts of Europe. At first, it was felt that the presence of haplogroup X in Native Americans was another example of recent (post-Columbus) admixture, but closer analysis suggests that haplogroup X in Native Americans predates European contact. Does this mean some European contact occurred before Columbus? In order to answer this question, it is necessary to look at the geographic distribution of mitochondrial haplogroups in Northeast Asia.

Although an Asian origin of Native Americans is definite, there is still debate about exactly *where* in Asia. The traditional explanation has been Siberia because of its geographic location and the close genetic affinity between Arctic Northeast Asians and Native Americans (Siberia is a large region that includes the Asian part of Russia and northern Kazakhstan). Siberian origins of Native Americans can be examined by looking at mitochondrial haplogroups in Siberian populations. If the first Americans came from Siberia and brought all four mitochondrial haplogroups with them, then we should find these same haplogroups in living populations in northern Siberia. The results are not entirely clear. Populations in northern Siberia do show haplogroups A, C, and D, but not B. However, some populations in southern Siberia, notably those around Lake Baikal (see Figure 6.4), do have haplogroup B, suggesting this region as a possible source for at least some of the migrants to the New World.

At first, the presence of the four haplogroups (A–D) in Siberia was seen as consistent with a Siberian origin of Native Americans. The finding of haplogroup X in Native Americans complicated the issue, particularly since X had been found in European populations but not in Siberian populations. What did this pattern indicate? One obvious possibility was that the presence of haplogroup X in the New World, but not in Siberia, meant that it was introduced from Europe. This does not necessarily mean genetic contact across the Atlantic. It is possible that haplogroup X could have come in with migrants across the Bering Strait that originated from the western parts of Eurasia, moving through Siberia en route to the New World but leaving no traces of haplogroup X behind, perhaps because of genetic drift. Either way, the lack of haplogroup X in Siberia suggested the strong possibility of more than one source of migrants to the New World.

The situation changed following DNA studies of additional samples across Asia. It turns out that haplogroup X *is* found in some Siberian populations. In 2001, Miroslava Derenko and her colleagues found haplogroup X in the Altai Mountain region near Lake Baikal in southern Siberia.[10] This new evidence shows all the mtDNA haplogroups associated with the first Americans (A, B, C, D, X) were likely to have been present in this region, suggesting that there could have been a single source of migrants to the New World. All of the other haplogroups found in some Native Americans, such as H and J, appear to have been introduced *after* Columbus.

Inspection of overall haplogroup frequencies is only part of the story. Some studies use more detailed inspection of the specific DNA sequences

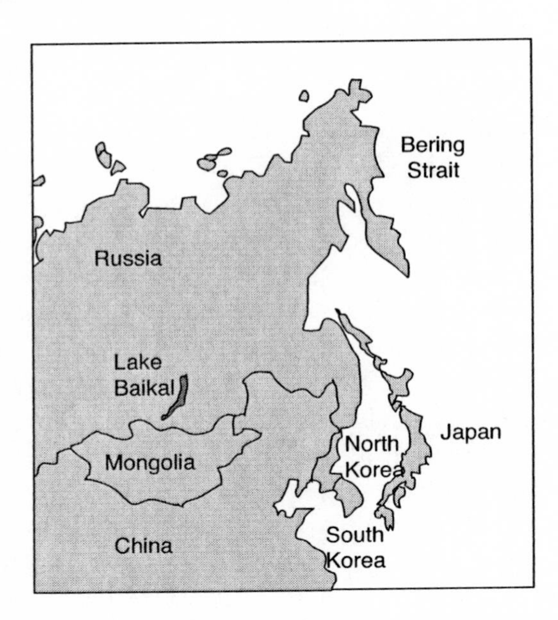

Figure 6.4 Location of Lake Baikal in southern Siberia. Mitochondrial DNA analysis suggests that the first Americans were descended from people who lived in this region, although some analyses point to the possibility of other sources of Asian migrants to the New World, such as eastern Asia.

found within the basic haplogroups (remember that a haplogroup is made up of a cluster of related haplotypes; although similar, they do not all have the exact same underlying DNA sequences). These sequence analyses suggest a more complex pattern of migration. Although haplogroups C and D are found throughout much of Asia, it turns out that specific mutations found in these haplogroups in Native American populations are found only in some *East* Asian populations, such as Japan and Korea, and not in the Siberian populations. On the other hand, the East Asian populations lack haplogroup B, which is found in Siberia. If there were more than one specific source of Asian genes, does this mean that there was more than one migration into the New World? Anthropologist Theodore Schurr of the University of Pennsylvania has argued that these analyses point to at *least* two separate migration events.[11]

How Many Migrations?

The question of the number of migrations into the New World remains contentious among both archaeologists and anthropological geneticists.[12]

Was there just a single migration event? If so, it could have lasted a ber of generations; remember that we are not talking about something that happened all at once. Were there several migration events, each separated by some interval of time? Some argue for a single migration, whereas others argue for two, three, or more. Early views tended to interpret population history in terms of supposed biological uniformity of Native Americans. It was felt that Native Americans were genetically homogeneous, which in turn suggested that there was a single migration to the New World, probably consisting of only a small number of founders.[13] As scientists came to realize that there was indeed more extensive genetic variation within the native populations of the New World, debate focused on the cause of this variation. Some argued that it reflected more than one migration, each bringing in a new combination of genes, whereas others felt that present-day diversity could be explained by a single migration followed by genetic divergence of American populations *after* they got into the New World.

In 1986, linguist Joseph Greenberg and biological anthropologists Christy Turner and Stephen Zegura argued for *three* separate migration events, based on linguistic, dental, and genetic evidence.[14] They started with Greenberg's suggestion that there are three different groupings of Native American languages—Amerind, Na-Dene, and Aleut-Eskimo—each of which has a distinct geographic distribution. Greenberg arrived at this classification after analyzing the similarity of words in different languages in an attempt to find out which ones were most similar to others, and whether there was a pattern to these relationships. According to Greenberg, the Amerind language group has the largest geographic distribution, including over half of North America and all of Central and South America. The Na-Dene language group is found primarily among natives of northwestern Canada but is also found in a section of the American Southwest. The Aleut-Eskimo group is found in Arctic North America.

They then looked for a correlation between this tripartite classification and the biological variation found among Native Americans. Their review of the dental evidence suggested the same basic pattern—three basic groups of Native Americans each defined by specific traits. The genetic data was also interpreted as supporting this tripartite classification. A rough correspondence between linguistics and genetics (including dental patterns) is not unexpected; both genetics and languages change because of differing rates of isolation and migration, and both reflect common ancestry to some extent. What needs to be explained is the *cause* for this three-part division of Native American populations. According to Greenberg, Turner, and Zegura, the most likely explanation is that each present-day language group represents the descendants of a separate migration into the Americas. The first migration consisted of the ancestors of the Amerinds, who traveled south through North America into Central and South America. The Na-Dene migration took place next, with descendants settling primarily in northwestern North America but with some moving into the American Southwest. The Aleut-Eskimo migration, settling into the northernmost parts of North America, came last (although they noted uncertainty about the chronological order of the last two migrations). Many linguists rejected Greenberg's tripartite classification, the cornerstone of their model. Others raised alternative interpretations of the biological evidence.[15] Although Greenberg's specific hypothesis has been rejected by many, the concept that there might have been more than one migration continues to receive some support.

The debate has sharpened somewhat since the rise of molecular genetics in the late 1980s. Analysis of classic genetic markers, such as blood groups, and of physical features, such as teeth and crania, are useful in determining population affinity, but analysis of DNA sequences can provide even more detailed information about population history because it becomes possible to track the specific history of mutations over time. However, there are still likely to be alternative explanations for any given observed pattern.

The evidence from mitochondrial DNA provides a good example of how diversity within Native Americans, and their relationship to Asian populations, has been interpreted in different ways. Some have argued that the presence of four distinct mtDNA haplogroups in North America corresponds to four distinct migration events.[16] Others maintain that there was only a single migration event because of shared mutations across the geographic range of Native Americans, suggesting a single origin and entry.[17] Some have also argued that diversity of mitochondrial DNA may not be able to resolve the debate over the number of migrations.[18]

Analysis of Y-chromosome haplotypes has provided additional insight on possible Asian source populations and the number of migrations. Initial work on the Y chromosome showed a single major haplotype found at high frequencies in a number of Native American populations, a finding that was suggested to reflect a single origin of Native Americans. More recently, however, more extensive surveys of Y-chromosome variation have supported a different scenario. Geneticist Tatiana Karafet a colleagues compiled a large global database of Y-chromosome types.[19] They found two Y-chromosome haplotypes with high frequency in Native American populations, both of which supported an Asian origin. However, the geographic distribution of these two haplotypes (labeled 1C and 1F) in Asia were quite different. Haplotype 1F had moderately high frequencies in groups around the Lake Baikal region, suggesting a likely source of the haplotype, and in agreement with some of the mitochondrial DNA evidence. On the other hand, haplotype 1C showed moderately high frequencies in western Siberia, suggesting a different source population. Karafet and her colleagues concluded that this evidence is best explained by *two* separate migration events from Asia, each from a slightly different region of Siberia.

When Was the New World First Inhabited?

When we consider the question of the number of migrations, we also need to consider the timing. When did this all take place? If there was a single migration, then when did it occur? On the other hand, if there were multiple migrations, when did they take place, in which order, and how much time separated them? Both supporters of a single migration and those supporting multiple migrations share one important question: When did the *first* Americans arrive?

For many years, the traditional archaeological answer has focused on a relatively "recent" first (or sole) entry, roughly 12,000 to 15,000 years ago. Numerous archaeological sites had been found throughout the Americas dating back to about 11,500 years ago, all sharing a particular stone tool culture known as the Clovis culture, named after a particular type of stone spear point first discovered at a site in Clovis, New Mexico. Allowing time for movement through the New World, these sites suggested that humans first entered the Americas more than 12,000 years ago, followed by a rapid spread of human populations across North, Central, and South America. The association of these tools with the bones of large animals, such as mammoths, suggested a widespread culture that relied extensively on hunting big game.[20]

Over the past few decades, there have been a number of claims for earlier occupation of the New World, with a number of sites being proposed as "pre-Clovis."[21] One of the oldest known sites, Monte Verde in southern Chile, is now thought to date back 12,500 years ago.[22] Because this site is about as far away from the Bering land bridge as one can get in the New World, and since it would have taken some time for even a rapidly expanding population to get that far, the early date for the Monte Verde site implies an earlier entry than usually suggested under the "Clovis first" model. There are a number of other suggested pre-Clovis sites, including Meadowcroft Rock Shelter in Pennsylvania, which appears to be older than 12,000 years, and the Cactus Hill Site in Virginia, which may be as old as 15,000 years, among others.[23] Although there is some skepticism regarding many of the early dates, careful analysis of the Monte Verde site has convinced many archaeologists that humans lived in the New World before the Clovis culture, thus pushing the minimum date for entry into the New World back even farther.

How much farther? Answering this question is a bit tricky because we need to estimate how fast the first populations could have expanded and spread across the New World. Given a date of 12,500 years ago for Monte Verde, it is obvious that the date of entry must be older because it would take time for migrants entering from Asia to disperse all the way south to the tip of South America. However, are we talking about a rapid rate of expansion, which would imply the first migration was not much earlier than 12,500 years ago, or a slower expansion, which would mean an even earlier date for the first migration?

We also have to consider physical barriers to movement. For example, if humans came via the Bering land bridge and then south through the ice-free corridor, then the possible dates of entry are limited to times when the sea levels were low enough to expose the land bridge and when the ice-free corridor was open. Many areas of possible movement were essentially blocked between about 13,000 and 20,000 years ago. Assuming that the first Americans came through the ice-free corridor means that inhabitation most likely took place more than 20,000 years ago or less than 13,000 years ago. If humans entered the New World 13,000 years ago, then they would have to have spread very quickly to reach Monte Verde by 12,500 years ago. If we consider a slower rate of expansion, then a date earlier than 20,000 years ago seems more reasonable. Of course, this argument rests on the assumption that the first Americans moved across the Bering land bridge. If further evidence confirms the suggestion that some humans used boats to enter the Americas, then we would have to consider a different set of variables and the possibility that entry occurred less than 20,000 years ago, but still sufficiently more than 12,500 years ago to allow time to reach Monte Verde. The continued testing of other suggested ancient sites could further change these numbers if the older ones are confirmed. At the moment, a broad estimate of between 15,000 and 20,000 years ago seems reasonable.

There have been claims to even older dates of entry to the New World. Some of these claims are based on archaeological evidence that has since been rejected, such as the suggestion by the late Louis Leakey that humans lived in California more than 50,000 years ago. Other claims for great antiquity have come from analyses of the genetic data, which some feel suggests an initial entry of 20,000 to 30,000 years ago.[24] These arguments are based on the estimated ages of the most recent common mitochondrial DNA ancestors in both Native American and Northeast Asian populations. Such estimates also figure into the argument over the number of migrations, with some evidence suggesting that haplogroup B might be younger and possibly indicating a separate migration event than that which brought haplogroups A, C, and D into the New World. Because there are many factors, such as population size, that can affect genetic estimates of the most recent common ancestors, it is not clear that a very ancient first migration is the only possible explanation. I suspect that the final determination of the age of the first Americans will be settled by archaeology and not by genetics. It is always possible that sites of equivalent age will be discovered and verified, but at present I doubt that the first entry of humans into the New World took place before 20,000 years ago.

Kennewick Man

Thus far, I have focused most of my discussion on the analyses of genetic data from *living* populations, but this is not the only way to reconstruct the past. Skeletal remains also provide a window into the past, either through analysis of various skeletal measures or, in some cases, extraction of mitochondrial DNA from skeletal remains. Such studies also confirm the Asian link. A recent and controversial find, known as "Kennewick Man," also shows the potential for scientific analysis of the first Americans and highlights the political and often contentious nature of prehistoric research in the Americas.[25]

[In] 1996, a human skull was found along the Columbia River in [Kennewic]k in Washington State. Archaeologist James Chatters examined the skull and concluded, based on facial and cranial features, that it was a man of European descent who was about forty-five to fifty years old when he died. Initially, there was no indication of anything unusual about the skull, and Chatters suggested that it was the remains of an old homesteader who had been buried on his farm.[26] However, the situation changed dramatically as more skeletal remains were found, including a stone projectile point embedded in the man's hip, which was *definitely* unusual since such points typically date back thousands of years, well before the arrival of Europeans in the Americas. This suspicion was confirmed when carbon-14 dating estimated that the man had lived and died somewhere between 9,200 and 9,600 years ago.[27] By itself, the date was not unusual since ancient skeletal remains of equivalent age have been found elsewhere. What *was* controversial, however, was the fact that the initial assessment of the skull showed European affinity thousands of years before Europeans had entered the New World.

Was this evidence of early European contact thousands of years before either Columbus or the Vikings (neither of whom made it to anywhere near the site of the find), or was Kennewick Man actually an early Native American who had been misclassified as European? The possible European connection was played up in the press following an initial reconstruction of the face, which suggested an appearance very similar to that of actor Patrick Stewart who plays, among other roles, Captain Jean-Luc Picard in *Star Trek: The Next Generation*. Others later suggested that the skull was actually more similar to living Native Americans than living Europeans, an explanation that does not require postulating ancient migrations from Europe that appear to have left no archaeological evidence.

At this point, the discovery of Kennewick Man started having both scientific and political implications. If Kennewick Man was Native American, then the skeletal remains are subject to the Native American Graves Protection and Repatriation Act (NAGPRA) of 1990, which requires repatriation of skeletal and cultural remains to the appropriate Native American tribe. On the other hand, if Kennewick Man represents an early European explorer or settler (regardless of how unlikely that seems), then NAGPRA is not relevant. The Army Corps of Engineers, which has jurisdiction over the land where Kennewick Man was found, concluded that the date of the specimen—in excess of 9,000 years, long before any documented European contact—meant that it was Native American and thus fell under the province of NAGPRA and took possession of the skeletal remains. Several local tribes, including the Umatilla, claimed that Kennewick Man was their specific ancestor and demanded the remains for reburial. Scientists wanting to study these remains, particularly in light of its age and presumed European features, argued against this and filed for permission to continue analysis.[28] Further complications were introduced when other groups also claimed Kennewick Man as their ancestor and wanted the remains turned over to them. One of these was the Asatru Folk Assembly, a religious group dating back to the time of the Vikings. If Kennewick Man was linked to Europeans, then they wanted the bones returned for appropriate reburial under Asatru tradition.[29] Various suits were filed as the story of Kennewick Man moved from the laboratory to the courtroom. Many of the legal arguments continue as of this writing. In mid-2002, a federal judge ruled in favor of scientists wishing to study the remains further, and appeals have since been filed by native groups as well as the U.S. government.

It now seems clear that Kennewick Man is not European. Preliminary comparative analyses of cranial and facial measures suggest that the initial assessment of European affinity was in error. There are several reasons for such misinterpretations. For one thing, populations overlap quite a bit for many cranial measures, and there is a general tendency for European and Native American crania to show this type of overlap, making it easy in some cases to misidentify skulls.[30] Another problem was that the initial assessment used reference measures from *living* populations for group classification of *ancient* remains. Comparing a 9,000-year-old skull with skulls from living populations does not take into account the fact that populations change over time. The oldest Native American remains tend to be somewhat different from those of recent Native Americans.[31] Living groups may not provide the best comparison. When Kennewick Man was compared to ancient crania from across the world, a different picture emerged. There is similarity to other ancient Native American crania, as well as similarity to some Asian populations.[32] As is the case with any study of population history over long periods of time, the picture is not always what we might expect, due to the continuing mixing and evolution of populations.

The legal battles are still ongoing and illustrate the conflict that often arises when there are opposing views about ownership of the past. Should these bones remain in the hands of scientists? Although they do offer the possibility for additional insight into the history of the first Americans, the belief of native groups that these bones are sacred and must be reburied needs to be respected. Although available evidence supports the idea that Kennewick Man was not a European but an early Native American, there is little evidence to link him with any *specific* tribal group in a genetic sense because of the constant mixing of genes between populations. We have a tendency to view the history of human populations as a series of independent lines connecting ancestors and descendants rather than the complex intermixing of different ancestral lines over time.

Past and Present

Molecular genetic data have confirmed the Asian origins of Native Americans suggested by earlier studies of blood groups, other genetic markers, and physical traits. There is less resolution on the question of the number of migrations or the timing of entry into the New World, both of which continue to be important research areas. Analyses of mitochondrial DNA and Y-chromosome DNA suggest close links between living Native Americans and the people living near Lake Baikal in Siberia, populations that show all four primary Native American haplogroups (A–D) as well as haplogroup X. It is therefore very tempting to proclaim these people as the descendants of the first migrants to the Americas.

Doing so is a bit risky because such a conclusion would be based on the assumption that there is a direct and unbroken line of descent from past to present that is independent of other populations. Living populations should not be viewed as the equivalent of "living fossils," species that have shown little change over long periods of time. The present-day genetic composition of the population residing in the Lake Baikal region in southern Siberia *does* contain valuable clues for reconstructing population history, but we should not fall into the trap of thinking of the present-day genetic diversity of these groups as necessarily identical to those of their ancestors.

Populations change genetically over time. Genetic drift can lead to deviations from previous generations. Gene flow between populations alters their genetic composition. Mutations continue to accumulate. The demographic structure of groups changes over time, and each change can have an impact on genetic diversity and genetic distance. Some populations expand in size, thus slowing the impact of genetic drift. Other populations undergo severe reductions in population size, perhaps due to rapid environmental shifts or colonization events, where only a small remnant of a population (and their genes) enters a new place. In such cases, the impact of genetic drift is likely to be greater. Groups fission into smaller groups, and sometimes small groups fuse back together. Mates are exchanged over both short distances and through long-distance movement. All of these potential events, and others, can alter the genetic composition of populations, making our efforts at reconstructing population history very difficult at times. These potential events become even more significant when we are trying to confirm very specific details, such as the number of migration events or the date of entry into the New World.

As an example, consider the mitochondrial DNA haplogroup distribution found in northeastern Siberia in those populations that today live *closest* to the Bering Strait. These populations, including the Chukchi and Siberian Eskimos, have haplogroups A, C, and D, but not B or X. On the other hand, populations residing near Lake Baikal in the *southern* part of Siberia, farther away from the Bering Strait, do have all five haplogroups. If the people living near Lake Baikal are indeed representative of the first migrants to the New World, and they moved through the Bering Strait, then why don't the populations living closest to the Bering Strait have haplogroups B or X? One possibility is that these haplogroups *were* present in these populations in earlier times but have since been lost through genetic drift. It is obvious that in this case we would not want to consider the present-day population living near the Bering Strait as genetically *identical* to those living there thousands of years ago. Genetic changes in Siberian populations, both in the north and south, likely altered the pattern of population relationships over time. We should therefore not rely so heavily on the assumption that the fine detail of *present-day* patterns of genetic variation was frozen in time.

We must be careful not to make too many inferences about earlier populations based on the genetic composition of living populations. Populations change over time, and the more time that elapses, the greater the difficulty of using living samples as proxies for earlier ones. If we are making inferences over relatively short periods, such as the examples from Ireland that I discuss in Chapter 9, then the danger of misinterpretation is slight. Over much longer intervals of time, such as the case of the Neandertals discussed in Chapter 4, the danger is much greater. We have an unfortunate tendency to think of our ancestry in terms of unbroken chains to the past, independent of other people's ancestry. Our conception of ancestry is generally limited to the generations we know most about, such as parents or grandparents. If all four of your grandparents came from Italy, you would characterize yourself as Italian-American. Does this mean that *their* grandparents all came from Italy? How about *their* grandparents? The farther we go back in time, the more mixed our ancestry becomes. We can use living populations as guides to population history, but we must always acknowledge the fact that all populations continue to change over time. Questions of population history and ancestry are often difficult to answer because the long-term evolution is more like a complex web of interconnections than a simple series of lines of descent. Kennewick Man appears to be Native American, but there is no reason to assume that he would be more similar to populations living in the same geographic area 9,000 years later than to other native groups living nearby or even farther away. The farther we go into the past, the more diffuse our ancestry becomes. General patterns of population history can be determined, but the more specific ones often elude us.

SEVEN

Prehistoric Europe: The Spread of Farming or the Spread of Farmers?

Do you ever think about how your ancestors lived? I recall being amazed that my grandparents did not have television while they were growing up (I've since stopped doing this, having to cope with my own children's astonishment that we did not have video games when I was a child, and my students' amazement that I did not have a word processor in graduate school). Comparisons across a few generations may seem quite large given the rapid pace of technological change during the past fifty years. Push it even further. Can you imagine what life was like a few hundred years ago? A few thousand? How about farther than that, pushing back before the existence of things we take for granted, like agriculture and complex societies?

One of the great educational benefits of anthropology is that it shows us that we live in a world today that is very different from that of our ancestors at the end of the last major ice age 10,000 years ago. Back then, humans subsisted primarily by hunting and gathering, obtaining their food directly from the environment. Our ancestors lived in small groups, generally interacting with only a couple of dozen people their entire lives. They were less likely to settle down in one area year-round but instead moved about in search of food, sometimes over wide areas and sometimes within more limited ranges. Moreover, their technology was quite different from ours. Biologically, the human species has not changed much over the past 10,000 years. Bones are slightly less rugged, and teeth are slightly smaller, but overall there has not been much change in the species as a whole. Ten thousand years is too short a time for major genetic shifts.

The cultural evolution of our species is another thing altogether. Human cultures can change at a much faster rate than human genes. In the past 10,000 years, we have gone from primary reliance on hunting and gathering to almost exclusive reliance on agriculture. Very few people in the world today rely on hunting and gathering as their major source of food. Everyone else relies on agriculture. The human species has increased dramatically in size, increasing from perhaps 5 to 10 million people worldwide at the end of the ice ages to more than 6 billion people today, a thousandfold increase. Our species continues to increase in size; it was estimated that our total numbers grew by 80 million in 2001, which works out to approximately 2.5 people per second.[1]

My point here is that culture change has been very rapid, progressing from hunting and gathering to agricultural societies in the blink of an evolutionary eye. Throughout human evolution, there have been several "revolutions," each of which has dramatically changed the future path of human history. The initial development of stone tool technology and the rise of hunting and gathering adaptations, starting about 2.5 million years ago, was one such revolution. Another is the Agricultural Revolution of the past 10,000 years, which permanently altered the basic structure of human societies. Due to changes associated with agriculture, we now live in large complex societies with formal political structures and social and economic stratification—a world far different from the one our recent hunter-gatherer ancestors knew.

Archaeologists have long been interested in the origin of agriculture. When, where, and why did it happen? Genetics has been able to provide insight into these events, particularly the rise of agriculture in prehistoric Europe, the major focus of this chapter. The archaeological record shows that agriculture spread into Europe from Southwest Asia (that region also known as the Middle East or the Near East). What has been less clear is the nature of this spread. Did agriculture spread as a cultural innovation from population to population from Southwest Asia into Europe, with new populations successively adopting this new way of living over time? Alternatively, populations of farmers might have moved out of Southwest Asia, perhaps in response to increased demand for land, and into Europe, mixing with indigenous hunting and gathering groups and carrying their new adaptation with them. In other words, what moved? Farming methods or the farmers themselves?

Origins of Agriculture

Before addressing this question, we need to put the origins of agriculture into broader perspective. Note that I use the plural form ("origins") rather than singular ("origin"), because it is clear from the archaeological record that there was no *single* initial origin of agriculture.[2] Instead, agriculture developed independently in different parts of the world and at different times during the past 10,000 years. This view is in contrast to a dominant view in nineteenth-century anthropology that tended to explain cultural behaviors as resulting from *diffusion,* the spread of ideas from place to place. The presumption of that view is that any new change, such as art or pyramid building or agriculture, must have arisen in one place and then spread out to other groups from there. The idea that a new idea or technology could have been independently invented was not popular at that time.

In contrast with this earlier view, current archaeological evidence shows that agriculture developed independently in several parts of the world. The primary centers of agriculture were located in parts of Southwest Asia, East Asia, sub-Saharan Africa, North America, Central America, and South America. The first evidence of the domestication of plants and animals dates to about 12,000 years ago, and rapid intensification of agriculture as a way of life then begins at different times in different places. In Southwest Asia, there is evidence of early domestication of dogs and sheep, followed by goats, wheat, barley, and lentils (among others) by 11,000 years ago. Agriculture arose independently in East Asia, with domestication of rice occurring roughly 7,000 to 8,000 years ago. The initial development of agriculture corresponds to a change in technology as stone tools were modified for use in harvesting grain and other agricultural tasks. The agricultural cultures of this time are often referred to as *Neolithic,* which translates as the "New Stone Age," in contrast to earlier stone tool cultures known as the *Paleolithic* and *Mesolithic.*

There have been many hypotheses suggested for the origins of agriculture. A common focus is climate change and population pressure. During the glacial times of the Pleistocene (the geological time period ending 10,000 years ago), the climate was simply too inhospitable for agriculture. The climate improved at the end of the Pleistocene as the glaciers receded and the global temperature increased. A changing climate, along with increased efficiency of hunting-gathering adaptations, led to growth in ɲulations. Humans also began settling down in year-round set-, able to make the most out of the improved conditions for hunt-d gathering. As populations grew, the need for further food rces also grew, stimulating a shift toward agriculture and away from unting and gathering. In turn, agriculture and a sedentary life led to higher levels of fertility, perhaps because of dietary changes leading to a shortened period of breastfeeding and hence shorter intervals between births. As populations continued to grow in size, the need for agriculture increased even more. Population pressure thus led to an increase in agriculture, which led to larger populations, which in turn led to further reliance on agriculture, round and round in a circle of cause and effect.

The preceding is, of course, an oversimplification of the process, and the dynamics of the relationship among agriculture, climate, and population growth varied from place to place, as well as available local resources and other historical contingencies. The details of the shift to agriculture remain a major research concern of archaeologists. The main point is that although we are still not completely clear on the details of causation, it is clear that agriculture developed independently in many different parts of the world.

The Origin of Agriculture in Europe

Agriculture did not develop independently in all parts of the world, however. Europe is the exception to the rule of independent origin. The archaeological evidence is quite clear that Europe represents a case of the diffusion of agriculture from Southwest Asia. The region known as the Fertile Crescent in Southwest Asia has the oldest documented agricultural sites. Before the rise of agriculture, people in this region relied primarily on hunting and then began increasing utilization of plant foods, which is evidenced by the appearance of stone tools, such as sickle blades and grinding stones, used to process plants. The remains of various cereal crops, such as wheat and barley, show evidence of domestication by about 11,000 years ago.

Following the origin and expansion of agriculture in Southwest Asia, we begin to see evidence of farming in southern Europe within the next few thousand years. With the exception of cattle (which appears to be European in origin), all domesticated plants and animals found in Europe appeared earlier in Southwest Asia, and it is generally acknowledged that

European agriculture was not developed independently but was the r
of diffusion.

The spread of agriculture began over 9,000 years ago, starting from the area that is now Iraq and Turkey and moving northwest into Europe. By 6,000 years ago, agriculture had reached the northwest corner of Europe. Archaeologist Albert Ammerman and geneticist Luca Cavalli-Sforza have calculated that agriculture spread at the average rate of roughly one kilometer per year in a northwest direction (Figure 7.1). Their spatial analysis of the history of European agriculture suggests a regular process of diffusion over time from Southwest Asia.[3] Their research was focused on the nature of this diffusion. Was it the knowledge of agricultural methods that spread over time and space, or was this diffusion associated with the migration of farming populations into Europe?

As an analogy, consider eleven people spread out across a football field, with one person at the goal line at each end of the field and everyone else spaced apart at 10-yard intervals. The person at one end has a football. How can the football get to the other end of the field? One way is for the person with the football to pass it to the next person 10 yards away, who in turn passes it to the next person, and so on, until the football has crossed the entire field. Note that the football moves, but the people do not. The other way the football can cross the field is if the first person walks across the entire field. Here, both the football and the person holding the football move across the field. In the case of the spread of agriculture in Europe, we want to know if the behavior spread from group to group, without the groups moving, or whether the spread of the behavior was accompanied by the actual movement of farming populations across Europe.

Cultural Diffusion Versus Demic Diffusion

Ammerman and Cavalli-Sforza noted that the diffusion of agriculture into Europe could have happened in these two different ways. The first model, known as *cultural diffusion,* proposes that farming methods and technology were behaviors that spread among indigenous groups, from village to village over time, analogous to passing the football from person to person. The people stayed where they were and did not move, but passed on new methods and knowledge about farming to neighboring groups. Hunting and gathering populations remained where they were

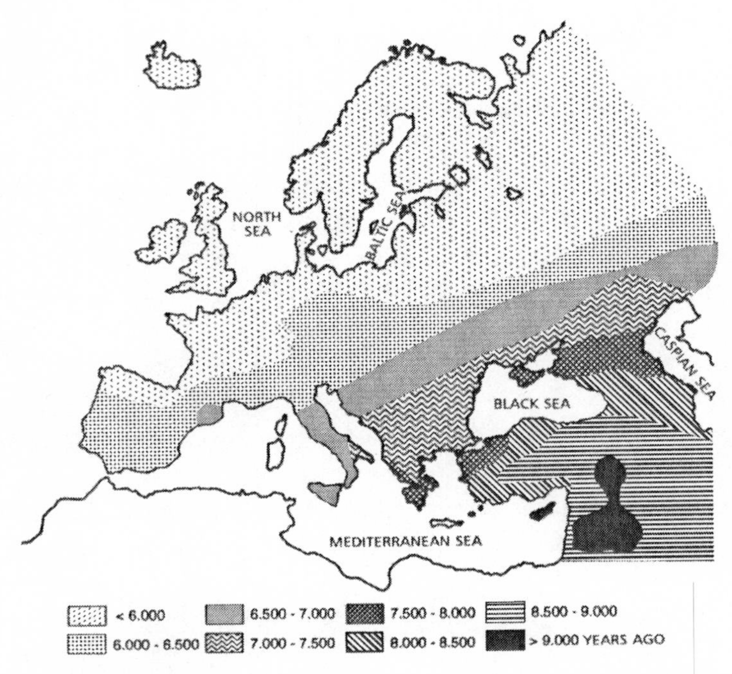

Figure 7.1 The origin of agriculture in Europe. This map shows the spread of agriculture out of Southwest Asia into Europe. The different graphic patterns correspond to the dates of the first appearance of agriculture throughout Europe based on carbon-14 dating of archaeological sites. There is a clear pattern of diffusion out of Southwest Asia in a northwest direction into Europe.
Source: L. L. Cavalli-Sforza and F. Cavalli-Sforza, *The Great Human Diasporas* (Cambridge: Perseus, 1995), Figure 6.5, p. 135.

but shifted to agriculture through the process of cultural contact and diffusion. In this way, farming spread into and throughout Europe without any actual movement of population.

Ammerman and Cavalli-Sforza questioned whether cultural diffusion was the cause of the spread of agriculture in Europe. They proposed an alternative model, known as *demic diffusion,* which focused on the movement of actual people, as opposed to behavior, across space (*demic* refers to *demes,* which are local breeding populations). Their model focused on the migration of people following population expansion. As agricultural populations in Southwest Asia expanded in size, they needed to expand in space as well to meet the resource needs of a growing population. As the farmers expanded, they physically moved farther and farther into Europe, where they genetically mixed with the indigenous hunter-gatherer populations. According to the demic diffusion model, the spread of agricu into Europe represents the movement over time of an expanding farming population.

How can we tell which of these models is correct? The archaeological evidence, which documents the actual spread of agriculture, could be interpreted both ways. If agriculture spread through cultural diffusion or through demic diffusion, the same pattern would result. Because both processes take time, both would produce the observed pattern of the movement over time from Southwest Asia into Northwest Europe. If all we had to rely on were the dates for the adoption of agriculture, we couldn't readily distinguish between the two diffusion models, because they both predict the same result.

Ammerman and Cavalli-Sforza cleverly realized that genetic data could help determine which of these models is appropriate. Under a pure model of cultural diffusion, only ideas and behaviors, but not genes, would move from population to population across all of Europe. The genetic diversity we see today in Europe would reflect primarily the underlying genetic variation present in the indigenous hunting and gathering populations before the spread of agriculture, since the primary change was cultural, not genetic. Under the demic diffusion model, both cultural behavior *and* genes would move over space, because when a population physically moves, so do the genes of the people. Under this model, genetic variation in living Europeans reflects the mixture of the expanding farming populations from the southeast and the indigenous hunting-gathering populations. Although the temporal distribution of evidence of first farming in Europe could be explained by either diffusion model, the genetic data had the potential to support one over the other.

In order to explain their proposal, Ammerman and Cavalli-Sforza created a simplified example, which I present here.[4] Start with three geographic regions: Southwest Asia, Eastern Europe, and Western Europe. The archaeological record shows that agriculture spread from Southwest Asia into Eastern Europe, and later into Western Europe. Imagine that prior to the spread of agriculture there was an allele that was found in Southwest Asia at a frequency of 100 percent, but that did not exist in Europe. Schematically, this situation would look like this:

Southwest Asia	Eastern Europe	Western Europe
100%	0%	0%

Under a pure cultural diffusion model, with no gene flow, these numbers would not change following the spread of agriculture (since this model posits the spread of ideas, not genes).

Now, consider what would happen under the demic diffusion model, where the population in Southwest Asia expands outward, first into Eastern Europe and then later into Western Europe, mixing with the indigenous hunting-gathering populations at each step. To illustrate this process, let's take one step at a time. First, some farmers expand out of Southwest Asia into Eastern Europe and mix to some extent with the indigenous hunter-gatherers in Eastern Europe. For the sake of argument, let's imagine a mixture rate of 50 percent, so that over time the population of Eastern Europe would have 50 percent ancestry from the indigenous hunter-gatherers and 50 percent from the expanding population of farmers. The allele frequency in Eastern Europe would now be equal to the average of the initial frequencies in Southwest Asia (100 percent) and Eastern Europe (0 percent), which gives an average of 50 percent. The allele frequency in Western Europe would stay the same because we have not yet gone to the next step, where the farmers in Eastern Europe further expand into Western Europe. At this intermediate point in time, the farmers have expanded only into Eastern Europe, and the allele frequencies would look like this:

Southwest Asia	Eastern Europe	Western Europe
100%	50%	0%

In the final step, the farmers in Eastern Europe now expand into Western Europe and mix with the indigenous groups living there. To make things simple, let's use the same mixture of 50:50. After this expansion, the allele frequency in Western Europe will be the average of the allele frequency of the incoming farmers from Eastern Europe (50 percent) and the allele frequency of the hunter-gatherers already living in Western Europe (0 percent). This average is 25 percent, and the allele frequencies now look like this:

Southwest Asia	Eastern Europe	Western Europe
100%	50%	25%

The end result is a *cline,* a geographic gradient in allele frequencies. The frequency of this hypothetical allele declines as we go from Southwest

Asia to Eastern Europe to Western Europe. Of course, the above example is overly simplistic and does not factor in other potential complications, such as gene flow between neighboring groups in addition to population expansion and mixing. The demic diffusion model is mathematically much more complex than shown here, but the overall interpretation is the same; demic diffusion will produce a cline, whereas a pure cultural diffusion model will not.

Genetic Evidence for Demic Diffusion

Cavalli-Sforza and his colleagues have presented genetic evidence for the demic diffusion model in a series of papers and books published since the late 1970s.[5] They initially noted that the spatial distribution of some genes fit the pattern expected under the demic diffusion model. A good example is the Rhesus blood group, which was discussed in Chapter 5. Figure 7.2 shows the distribution in Europe of individuals who have the Rh– blood type. Populations in Southwest Asia tend to have a fairly low percentage of individuals with Rh– blood, but this percentage increases in a northwestern gradient into Europe, where the percentage of those with Rh– blood is higher. This overall picture corresponds nicely with the prediction of the demic diffusion model, although even here there are exceptions, such as the much higher occurrence of Rh– blood in populations in northern Spain and southern France, associated with the Basque people living in that region.

Other genes showed the pattern predicted by demic diffusion, but not all. Some showed clines, but not in the expected northwest direction out of Southwest Asia. An example of one of these other genes is the frequency of the *B* allele for the ABO blood group, shown in Figure 7.3. Here, the frequency of *B* is highest in southern Russia and declines in an east-west gradient rather than a southeast to northwest direction. Although this definitely is a cline, it appears not to be related to the expansion of farmers out of Southwest Asia. What could account for this? It is possible that this cline is the result of a different population expansion from the east.

Rather than trying to make interpretations on a gene-by-gene basis, Cavalli-Sforza and his colleagues reasoned that it made more sense to use a method that allowed for the *simultaneous* comparison of spatial patterns in a large number of alleles from different genes. Migration should affect

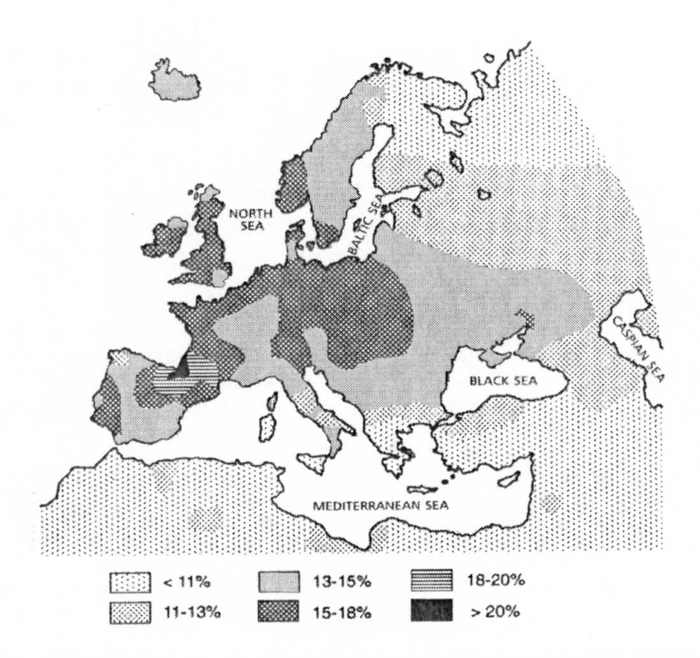

Figure 7.2 The spatial distribution in Europe of individuals with the Rh– blood type. The frequency of people with Rh– blood increases progressively in a northwestern direction from Southwest Asia to Western Europe, although the maximum lies in the Iberian Peninsula. This picture is somewhat consistent with the spread of agriculture.
Source: L. L. Cavalli-Sforza and F. Cavalli-Sforza, *The Great Human Diasporas* (Cambridge: Perseus, 1995), Fig. 6.9, p. 145.

all genes, and a method that allowed comparison across many genes would provide the best test of the demic diffusion model. The method they settled on is a statistical technique known as *principal components analysis.* This method allows the researcher to find common patterns of correlation among the frequencies of different alleles in various European populations.

Imagine a large table where the rows correspond to various locations in Europe and the columns correspond to different alleles. Given a large number of locations and alleles, one could stare at this table all day long and not easily perceive common and unique spatial patterns. Principal components analysis provides a mathematical way of looking for common and unique patterns by examining how closely correlated different allele frequencies are, that is, it identifies similar spatial patterns. Principal com-

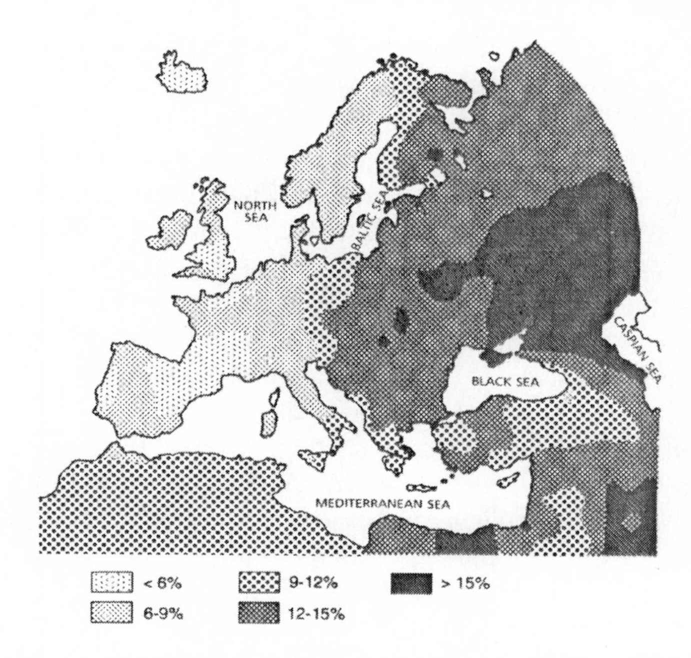

Figure 7.3 The spatial distribution in Europe of frequencies of the *B* allele for the ABO blood group. The frequency decreases from east to west, suggesting movements of people in the same direction. This pattern suggests an ancient expansion out of Eastern Europe rather than the expansion out of Southwest Asia associated with the spread of agriculture.
Source: L. L. Cavalli-Sforza and F. Cavalli-Sforza, *The Great Human Diasporas* (Cambridge: Perseus, 1995), Figure 6.9, p. 146.

ponents analysis starts by looking for the strongest common spatial pattern, known as the *first principal component,* and determines how much each allele contributes to that pattern. A numeric value is computed for each population corresponding to this first principal component. Once this is done, the method then identifies the *next* most common pattern among the variation that remains after the first principal component is extracted. This second principal component describes spatial variation that is unrelated to the first component; that is, it is left over after we extract the first principal component. Principal components analysis then identifies the third principal component (unrelated to the first two), the fourth (unrelated to the first three), and so on until most of the variation among alleles and populations has been accounted for. This method is useful in "peeling" away layers of genetic change, each of which might be associated with a different set of historical events.

Although the computations involved in principal components analysis are mathematically complex, the results provide an easy and useful way of testing historical explanations of genetic variation, because they can identify common spatial patterns across several different genes. Because there may be more than one underlying spatial pattern, the method produces more than one principal component. For each component, scores are computed for each population and then plotted on a map to allow a simple visual image of genetic variation across all alleles simultaneously. Once again, a picture is worth a thousand words.

The most recent analyses conducted by Cavalli-Sforza and his colleagues are based on ninety-five different alleles across a large number of European populations.[6] Figure 7.4 is a map of the first principal component (the one explaining the most variation among populations and identifying the major spatial trend). The actual scores of the first principal component are not shown, since that is too difficult to represent or interpret. Instead, the individual scores are collapsed into one of eight different intervals, each represented by a different graphic pattern (similar to the way temperature maps are shown in weather reports).

The first principal component (Figure 7.4) from Cavalli-Sforza's analysis shows a very clear pattern. The component scores change from a value of 8 in Southwest Asia to a value of 1 in Northwest Europe. (It doesn't matter which end has 1 or 8 since the direction and scale is arbitrary; what matters is the pattern of change over space.) There is a regular spatial pattern, showing a change in component score along a geographic gradient running in a southeast to northwest cline. This type of gradient is best explained by the expansion of a population with subsequent genetic mixing with indigenous populations. In which direction did this happen? The pattern in Figure 7.4 could be explained in one of two ways: There was either an expansion from Northwest Europe over time into Southwest Asia or an expansion starting in Southwest Asia and terminating in Northwest Europe. Given what we know about the spread of agriculture in Europe from carbon-14 dating (Figure 7.1), the second hypothesis—expansion from Southwest Asia into Europe—makes more sense. The first principal component runs in a northwest direction paralleling the movement of agriculture out of Southwest Asia. This correlation supports the demic diffusion hypothesis and shows that the major historical effect

from incoming Neolithic farmers, while populations farther away from Southwest Asia, such as France and Germany, have lower rates. This geographic pattern makes perfect sense under the demic diffusion model. As populations expanded out of Southwest Asia, those closest would have received a greater proportion of the genes from this expanding population.

What Else Was Happening in Europe?

There is growing evidence that the demic diffusion resulted in a mixture of genes from Paleolithic hunter-gatherers and Neolithic farmers expanding out of Southwest Asia. Is recent European genetic history that simple, or did other historical events also affect genetic variation in living Europeans? Although Cavalli-Sforza and his colleagues showed that there was a significant genetic impact corresponding to demic diffusion, this does not mean that this was the *only* such event that shaped European genetic diversity. Remember that the first principal component represents the primary source of variation, but not the only one. In their analysis, Cavalli-Sforza and his colleagues found that the first principal component accounted for 28 percent of the total genetic variation. This leaves 72 percent of the total variation that is *not* accounted for by demic diffusion. What do these other principal components tell us about causes of genetic variation in Europe? No one claims that demic diffusion is the *only* thing that shaped European genetic variation. Clues for other factors can be found by examining some of the other principal components.

Figure 7.5 shows the second principal component, which accounts for 22 percent of the total variation. There is a definite north-south pattern of variation; one extreme is in the Iberian Peninsula (Spain, Portugal, and Gibraltar), and the other is in northernmost Europe. This pattern could represent an expansion out of Northern Europe southward or an expansion out of the Iberian Peninsula northward. Based on research led by Antonio Torroni,[10] I find the second hypothesis more likely. Torroni and his colleagues found evidence of a south to north expansion based on European mitochondrial DNA. They found that haplogroup V provides evidence for an expansion out of the Iberian Peninsula between 10,000 and 15,000 years ago. This expansion might reflect the movement of people out of an isolated area after the end of the last glacial maximum. According to this view, some populations in the Iberian Peninsula were isolated during the most severe recent ice ages and then moved outward

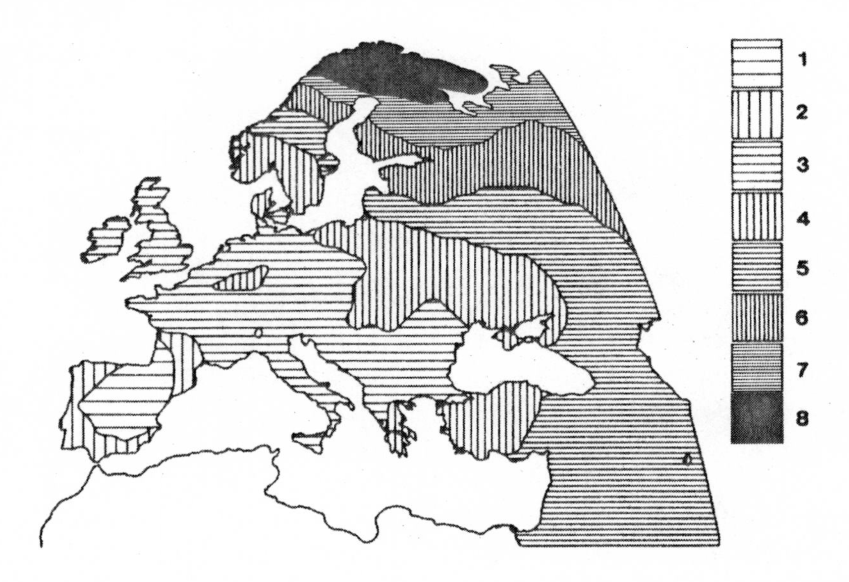

Figure 7.5 Genetic evidence for an expansion out of the Iberian Peninsula? This map shows the spatial distribution of the second principal component of genetic variation based on 95 allele frequencies. The legend box shows the order of principal component scores from one extreme to the other. This map shows a gradient northward from the Iberian Peninsula, which may correspond to an ancient movement between 10,000 and 15,000 years ago, perhaps as humans expanded back into parts of Europe following the last glacial maximum.
Source: L. L. Cavalli-Sforza and F. Cavalli-Sforza, *The Great Human Diasporas* (Cambridge: Perseus, 1995), Figure 6.11, p. 154.

following the end of the last ice age to mix with nearby populations across Europe. A post-glacial expansion makes sense when we think about the likely impact of climate on migration. During glacial times, population movements may have been more restricted. When the climate warmed and conditions improved, these previously isolated populations would have been free to expand again. Perhaps this is what is being picked up by the second principal component in Cavalli-Sforza's analysis.

What else can we detect? Each subsequent principal component accounts for a smaller portion of the residual variation. Since the first component accounts for 28 percent of the total variation and the second component accounts for 22 percent of the total variation, this means that the first two principal components collectively account for 50 percent of the total variation. This still leaves 50 percent. Are there any genetic signals buried in this residual variation? We answer this question by turning to the next (third) principal component, shown in Figure 7.6, which accounts for 11 percent

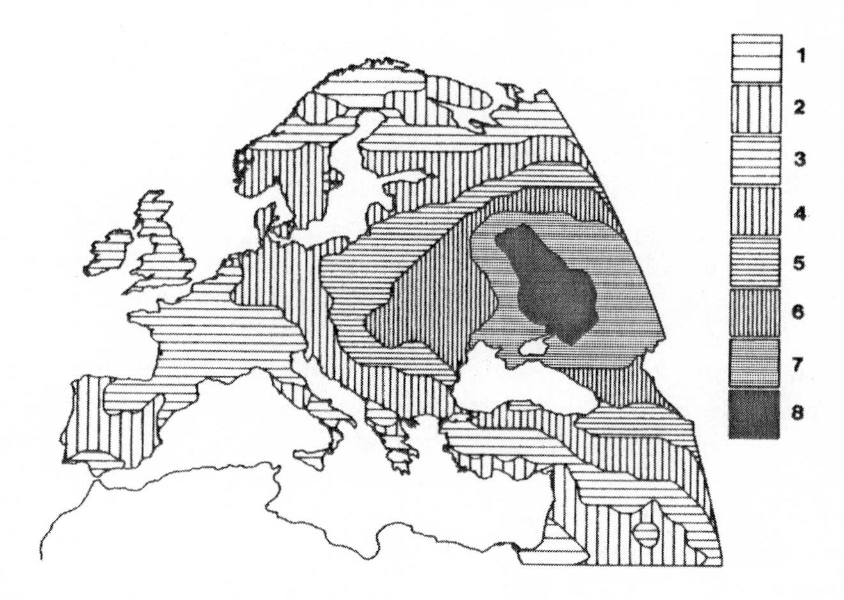

Figure 7.6 Genetic evidence for the expansion of the Kurgan people? This map shows the spatial distribution of the third principal component of genetic variation based on 95 allele frequencies. The legend box shows the order of principal component scores from one extreme to the other. This map shows an east-west gradient centered in southern Russia, which may correspond to waves of migration from the Kurgan peoples of that region.
Source: L. L. Cavalli-Sforza and F. Cavalli-Sforza, *The Great Human Diasporas* (Cambridge: Perseus, 1995), Figure 6.12, p. 155.

of the total genetic variation. The third principal component shows a different spatial pattern. Instead of a simple cline, there is a concentric pattern radiating outward from the Black Sea. This pattern may correlate with the expansion of the Kurgan peoples of southern Russia into the rest of Europe that started around 6,000 years ago, perhaps as a result of their domestication of horses.[11]

Many linguists have suggested that the Kurgan culture may be the initial source of the Indo-European language group, to which most present-day European languages belong. Linguist Marija Gimbutas has suggested that Indo-European speakers first entered Europe in three westward waves of migration. If so, then the third principal component may be picking up the genetic effects of these migrations. An alternative view has been proposed by archaeologist Colin Renfrew, who proposes that Indo-European languages spread out of Asia Minor as part of the expansion of Neolithic farmers. Can the principal components analysis be used to distinguish between these two hypotheses? No, because all we can see from these analyses is genetic evidence of the Neolithic expansion (the first principal component) and of the Kurgan expansion (the third principal component). We can't tell which of these was associated with the spread of Indo-European languages.[12] To do this, we need to have some way of correlating genetic changes with linguistic changes.

Evolutionary biologist Robert Sokal and his colleagues developed a way to do this.[13] They looked at "linguistic distances," which are measures of differences among modern European languages, and examined how strongly they correlated with genetic distances among the same set of populations. Both the Gimbutas hypothesis and the Renfrew hypothesis provide an explanation for the correlation between genetic and linguistic distance; one proposes westward expansion of the Kurgan, whereas the other proposes northwest expansion out of Southwest Asia. Sokal and his colleagues then figured out what the genetic distances would look like if Gimbutas's model were correct and tested to see if these expected distances provided a better "fit" to the observed genetic distances than the Renfrew model. Then they did the same thing with a set of expected distances based on Renfrew's model. Although this type of method can be useful in resolving historical hypotheses, that did not happen here; they were not able to show which hypothesis, if either, was most appropriate. To date, genetic data have not been able to resolve the question of the origins of Indo-European languages.

Getting back to the principal components analysis, we see that the first three principal components collectively account for 61 percent of the total genetic variation (28 + 22 + 11 = 61). Each subsequent principal component accounts for progressively less of the genetic variation but may represent historical events that, though smaller in magnitude, nonetheless affected genetic variation. These patterns can provide clues to population history. For example, the fourth principal component (Figure 7.7) accounts for only 7 percent of the total variation but shows an interesting pattern. Figure 7.7 shows a series of concentric rings radiating outward from Greece. This pattern may reflect expansion of the Greek empire in historical times (although the reason for the reversal in trend in northern Scandinavia is not clear).

The fifth principal component (Figure 7.8) accounts for only 5 percent of the total genetic variation but shows a very interesting localized pattern. There is a steep gradient around the Iberian Peninsula, with extreme com-

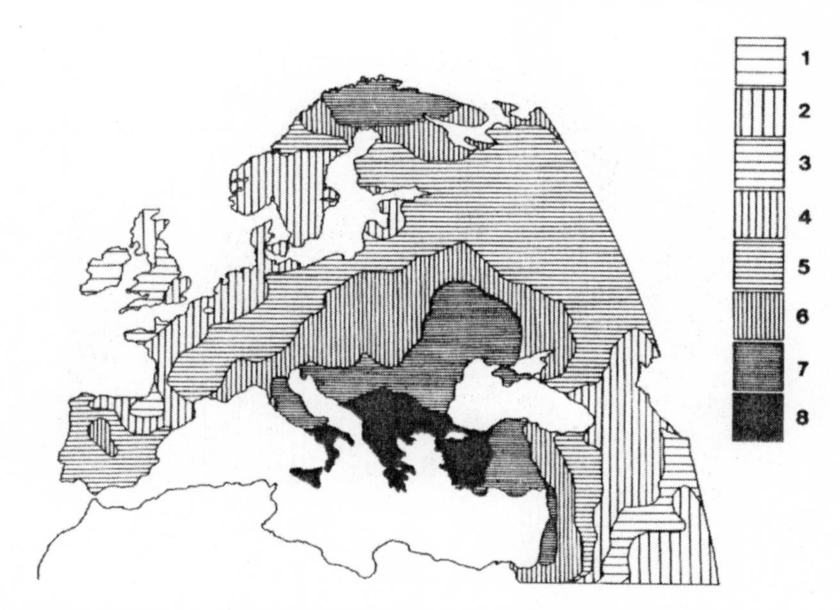

Figure 7.7 Genetic evidence for expansion of the Greek Empire? This map shows the spatial distribution of the fourth principal component of genetic variation based on 95 allele frequencies. The legend box shows the order of principal component scores from one extreme to the other. This map shows a gradient that may correspond to expansion of the Greek Empire.
Source: L. L. Cavalli-Sforza and F. Cavalli-Sforza, *The Great Human Diasporas* (Cambridge: Perseus, 1995), Figure 6.13, p. 156.

ponent scores (marked in black) in the region where the Basque language was once spoken (the language has a somewhat smaller range today). The Basque population is divergent linguistically and genetically from other western European populations and has remained relatively isolated from the rest of the world. The Basque language is one of the few non-Indo-European languages in Europe, and its origins are still unknown. Genetically, the Basque population tends to be dissimilar to their neighbors, and various hypotheses have been proposed to determine how much of this difference reflects the genetic effects of cultural isolation and how much reflects population history. Although gradients in principal component maps are often associated with population expansions (true for the first four principal components), this does not appear to be the case for the fifth principal component, given what is known about European history. Instead, it appears that the Basques are a relic population, perhaps dating back to very distant times, that has resisted both cultural and genetic mixture.[14]

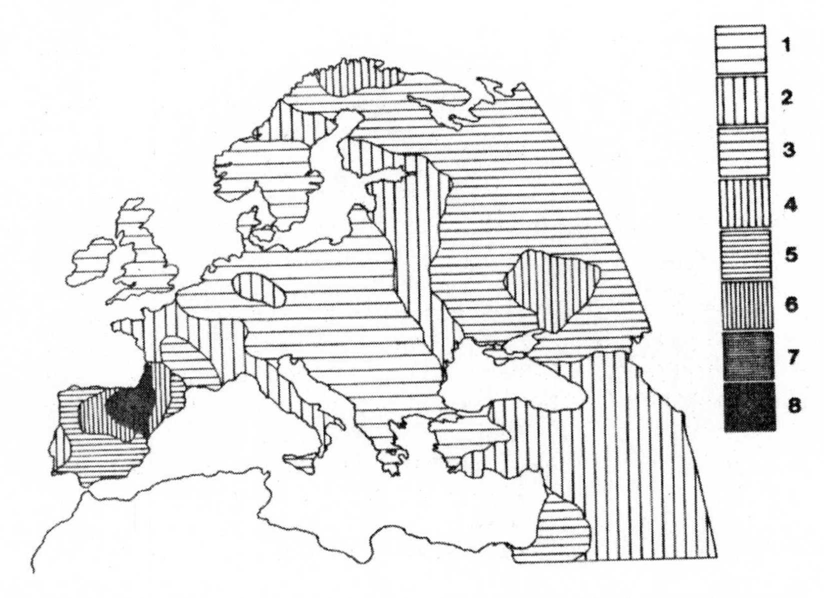

Figure 7.8 Genetic evidence for isolation of the Basques? This map shows the spatial distribution of the fifth principal component of genetic variation based on 95 allele frequencies. The legend box shows the order of principal component scores from one extreme to the other. This map shows noticeable differences in the region where Basque speakers have lived and may represent the genetic effect of the cultural isolation of the Basque population.
Source: L. L. Cavalli-Sforza and F. Cavalli-Sforza, *The Great Human Diasporas* (Cambridge: Perseus, 1995), Figure 6.14, p. 157.

Peeling Away the Layers

In Chapter 5, I used the analogy of a palimpsest to describe population history. We can consider the current patterns of genetic variation as the end result of many past events, each of which overwrites some of the impact of previous events. Principal components analysis, combined with other genetic analyses, provides us with a way of peeling away the genetic layers representing different events in European population history.

We have seen that a major influence on European genetic variation was the movement of farming populations out of Southwest Asia, bringing both new technology and new genes into Europe. Almost as significant was the expansion out of the Iberian Peninsula before the onset of agriculture, likely reflecting the end of an ice age. The Kurgan expansion, a possible Greek expansion, and the isolation of the Basques have also shaped the genetic landscape of Europe. However, there is probably additional text hidden in this particular palimpsest.

When you look at any picture or object you can see both broad patterns and fine details. The patterns that have been discussed here belong to the broad picture as seen when viewing the European continent as a whole. No one claims that the expansions described here were the only population movements that occurred in Europe's past. We know from both archaeology and written history that there were many examples of migrations that took place on a more localized level. Many studies of European population genetics have helped round out our broad understanding by looking at the fine details.

One good example is the effort by Robert Sokal and his colleagues to construct a massive database consisting of ethnohistoric records of population movements in Europe over the past 4,200 years.[15] They collected information from a variety of sources on historically known movements that were due to settlement, invasions, expansions, and other migrations. Given the relationship between population movement and gene flow, we would expect that two populations known to be linked through migration would be genetically similar to each other, all other things being equal. Sokal and his colleagues have found this to be the case; historical movements correlate strongly with the genetic distance between pairs of European populations.

Another fine-scaled approach has been taken by Guido Barbujani and Robert Sokal in their analysis of zones of genetic change in Europe.[16] They looked at the spatial distribution of genetic distances, searching for examples where there is a sharp difference between geographically proximate populations. Normally, we would expect geographically proximate populations to be genetically similar, based on the assumption that geographic proximity increases the level of gene flow between populations and hence increases genetic similarity. Barbujani and Sokal found numerous examples of pairs of populations that were less genetically similar to each other than would be expected on the basis of geographic distance. They analyzed genetic distances using a method known as *wombling*, named after W. H. Womble, to detect places in Europe where there is a relatively abrupt genetic change from one population to another, of a magnitude greater than expected on the basis of geographic distance. They surveyed data on 63 alleles in more than 3,000 locations in Europe and found thirty-three cases of abrupt genetic change.

What could cause this discontinuity? There must be some physical or cultural barrier to gene flow. The answer here is a cultural barrier; the overwhelming majority (thirty-one) of these discontinuities corresponded to language boundaries, such as the one between Celtic and Germanic languages in northwestern Europe. Because language differences are indicative of cultural differences, we are probably seeing the genetic effect of cultural isolation. It is thought that the different languages in Europe probably reflect distinct groups of people moving in from distant areas, and the linguistic-cultural differences acted to inhibit gene flow among these groups to a marked extent.

When viewed from either a broad or a local perspective, it is clear that the genetic variation of living Europeans reflects a mosaic of past events that occurred over a long span of time. The study of European genetic history shows us that some groups expanded both numerically and geographically, while other groups were absorbed to various degrees. Some groups (for example, the Basques) remained more isolated and have persisted in their uniqueness to the present day. Language differences inhibit gene flow and act to maintain or even enhance genetic differences. At the same time, new migrations and local gene flow act to remove previous traces of genetic history. The genetic landscape is constantly shifting.

EIGHT

Voyagers of the Pacific

"Space—the final frontier." Most of us are familiar with this phrase, which began each episode of the television series *Star Trek* and *Star Trek: The Next Generation.* This image resonated with many of my generation who witnessed the early events of the dawning space age, including the historic voyages of Alan Shepard and John Glenn and the moon landing in 1969. The final frontier was being broached in both fiction and real life. By this point in human history, humans had permanently settled across the globe in all continents except Antarctica. The entire history of human evolution has been one of expansion and colonization of new frontiers. Early humans *(Homo erectus)* expanded out of Africa into parts of Asia and Europe. Archaic humans expanded farther into Europe. Modern humans first moved into Australia at least 60,000 years ago and into the Americas 15,000 years ago (or earlier). Technological progress and human problem-solving even allowed us to live in Antarctica and deep under the sea. Today, outer space does appear to be our current "final frontier," one that poses immense technological challenges and that is likely to be expensive in terms of money, time, and lives (as all new frontiers are).

As we ponder the implications of this impending phase in human history, it is useful to remember that our current efforts to conquer outer space are but the most recent in a long series of challenges posed by expansion and movement into different environments. The initial expansion out of Africa by *Homo erectus* should not be thought of as an easy stroll or a simple migration of people across space. It involved considerable challenges in adapting to new environments. Likewise, the initial occupation of Australia by humans 60,000 years ago represents a major feat, given the level of technology available at the time, as did movement into the Americas, which required adapting to environments as varied as arctic tundra and rainforest jungle. Throughout human evolution, the ability of our species to solve problems and pass information from generation to generation allowed our ancestors to push past the barriers and move on to each new frontier.

Before our quest to colonize outer space, the most recent human geographic expansion and settlement into a new frontier took place in the part of the Pacific Islands known as Polynesia (an appropriate name that translates from Greek as "many islands"). Polynesia is made up of many small islands in the Pacific Ocean that are located in a rough triangle made up by the three points of New Zealand, the Hawaiian Islands, and Easter Island off the coast of South America (Figure 8.1). Other Pacific Islands are geographically closer to the Asian mainland and include the islands making up Melanesia and Micronesia. Melanesia consists of New Guinea and several islands to its east, including the Solomon Islands and Fiji. The word "Melanesia" translates as "dark islands," so named because of the dark skin of native Melanesians. Micronesia ("small islands") is made up of more than 2,000 islands, including the Marshall Islands, Kiribati, and the Caroline Islands, all of which lie north of Melanesia.

This chapter examines the origins and population history of Polynesians. Parts of the Pacific Islands have long been settled by modern humans. Humans moved into parts of Melanesia at least 35,000 years ago from Australia, which in turn has been occupied for at least 60,000 years. Polynesia, however, is a much more recent frontier in human history, first settled by humans about 3,500 years ago. The large geographic range of Polynesians fascinated many European explorers who found natives living on islands throughout this area, ranging from New Zealand in the south to the Easter Islands to the east and as far north as the Hawaiian Islands. The wide geographic range across miles and miles of ocean posed a series of questions about their origins and their seafaring abilities. Where did they come from? How did they get there? The conquest of the Pacific Ocean by early Polynesians in their canoes is a remarkable achievement and represents the most recent case of humans settling in a new area on our planet. If space is our current "final frontier," one of the last previous ones was Polynesia.

Where Did the Polynesians Come From?

Linguistics provides some clues for Polynesian origins. There is considerable linguistic diversity among the native peoples of Australia and the

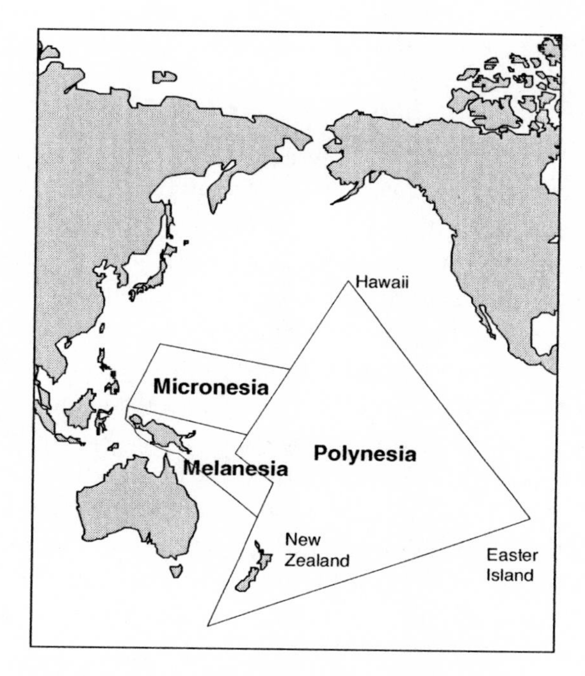

Figure 8.1 Locations of Micronesia, Melanesia, and Polynesia. Based on Lum et al. (1998).

Pacific Islands. Polynesian languages belong to a language family known as Austronesian, which includes almost 1,000 different languages in populations in both the Pacific and Indian Oceans, from Easter Island to Madagascar, but which differs from the languages spoken by native Australians and New Guineans. Austronesian speakers include inhabitants of Polynesia, Micronesia, and parts of Melanesia. Archaeological and linguistic evidence suggests that Polynesia was first inhabited about 3,500 years ago by Austronesian-speaking farmers who had expanded out of Taiwan or southern China.

How can we trace the Austronesian expansion into Polynesia? One source of archaeological evidence is pottery. Differences in the style and manufacture of pottery can be used to trace cultural connections over time and space. Archaeologists have found that the expansion can be traced by the presence of a particular form of pottery known as Lapita, characterized by a combination of horizontal bands, incised decorations, and geometric designs. We know more about Lapita culture than just their style of pottery; these people were farmers who also relied extensively on fishing for subsistence, and they used shells for both ornaments and tools. The Lapita culture is generally acknowledged to have been the forerunner of later Polynesian culture.[1]

When archaeologists plot the dates associated with the first appearance of the Lapita culture on a map, we see an interesting pattern that documents the rapid spread of Austronesian farmers eastward into the Pacific Ocean. The Lapita culture appeared first in the Bismarck Archipelago (a group of small islands slightly north of New Guinea) about 3,500 years ago and then moved eastward, reaching Samoa and the Cook Islands about a thousand years later, and then the Hawaiian Islands and Easter Island around 1,600 years ago. But where did the Lapita culture originate? Some archaeologists suggest that the Lapita culture is related to a similar one that existed in Taiwan and South China about 6,000 years ago, thus suggesting an origin in either East Asia or Southeast Asia (Southeast Asia is a region that includes the southeastern coast of mainland Asia as well as the island populations of Malaysia and Indonesia).[2]

The rapid expansion of the Lapita culture throughout the Pacific Islands is quite impressive, given the distances across water that were involved. All evidence points to these voyages being intentional and not simply the results of getting lost while out fishing. The voyagers took plants and animals with them as part of deliberate colonization efforts.[3] To do this they would have needed some basic knowledge of navigation. They relied on star patterns and observations of migrating birds, among other clues, to navigate without instruments. Their ability to travel across the ocean is even more impressive when you stop to consider that they did so in canoes! Think about the difficulty in traveling a long distance in a canoe, subject to a churning ocean. If you've ever used a canoe, you can imagine how hard it must be to keep your balance on an ocean (I have had trouble even on a calm lake). However, these were not the typical canoes with which you are probably familiar. It seems likely that these early ocean voyagers used double-outrigger canoes similar to those still used today by some Polynesian peoples. A double-outrigger canoe consists of a dugout canoe with two logs known as outriggers that are attached parallel to the main canoe. This arrangement provides additional buoyancy and prevents the canoe from rocking too much and tipping over.[4]

As noted above, archaeologists and linguists have traditionally pointed to Taiwan or the southern coast of China as the initial point of origin of the Austronesian expansion that populated Polynesia. From here, the expansion moved into the Philippines and Indonesia and then eastward into the Pacific Ocean.[5] One of the major questions is what happened along the way culturally and genetically. All evidence points to an expansion that went through part of Melanesia, including places that had already been inhabited for tens of thousands of years. It seems almost inevitable that the expansion of Austronesian farmers would have involved some contact with indigenous Melanesian peoples. How much? Did Melanesians contribute, either culturally or genetically, to Polynesians, and if so, then to what extent? Or did the Austronesian expansion pass them by?

There has been considerable debate about the nature and extent of contact with Melanesians. At one extreme is the "express train" model formulated by Jared Diamond, who proposes that the movement out of Asia was very rapid, so that the expanding Austronesian farmers essentially bypassed the Papuan-speaking peoples of Melanesia on their path eastward.[6] This model proposes little, if any, contact between the two groups. The "express train" analogy is apt; the very name conjures up an image of rapid movement with no stops along the way. The Austronesians zoomed past Melanesia. Diamond's model is in agreement with the earlier observations of Jules Sébastien César Dumont d'Urville, a noted nineteenth-century French explorer of Antarctica and the South Pacific. Dumont d'Urville is credited with the first use of the classification of Pacific peoples into the three groups of Micronesians, Melanesians, and Polynesians. Noting that the dark skin of the Melanesians was quite different from the lighter skin color of Polynesians, Dumont d'Urville concluded that these populations had no historical connection.[7] In other words, the express train roared past Melanesia straight into Polynesia with no genetic mixing along the way.

Is it reasonable to assume that no contact took place between these peoples? Many archaeologists think not and argue that the express train model is too extreme. Although the movement into the Pacific was rapid in terms of evolutionary time, it was not so fast that there would have been no opportunity for cultural and genetic contact along the way. Whereas the express train model argues for a distinct and recent Asian origin, others suggest that a model with some cultural and genetic mixture between an expanding population and the indigenous population of Melanesia is more appropriate.[8]

There is a third hypothesis of Polynesian origins—one that has never been popular among academics but did command considerable public attention for a time. This idea came from the Norwegian explorer Thor Heyerdahl, who suggested a New World origin for Polynesians. Heyerdahl noted that there are a number of linguistic and archaeological similarities between Polynesians and South American Indian peoples, similarities he believed could only be explained by the existence of a historical connection between the two populations. Therefore, he proposed that Polynesians came from South America instead of Asia. The basic assumption here is that any cultural trait found in two parts of the world must be connected through diffusion. Of course, similar characteristics could also be explained by independent invention and do not necessarily prove a historical connection (a classic example is the fact that both ancient Egyptians and the New World Maya built pyramids—a superficial similarity that has nothing to do with a historical connection).

Heyerdahl supported his model by showing that early South Americans *could* have made the voyage. To do this, he built and sailed a balsa raft, which he named the *Kon-Tiki,* from the coast of Peru to the Tuamotu Islands in the Pacific. Although he was successful in making this voyage using a level of technology equivalent to what would have been available to prehistoric Indians, it is not clear what this proved other than that it was *possible* to sail to Polynesia from South America. Proving something is possible is not the same thing as proving that something actually occurred. Most anthropologists have long rejected Heyerdahl's views.[9] Genetics has also supported this conclusion.

Genetic Distances and Polynesia

What do genetic data say about Polynesian origins? To start with, we can rule out a New World origin. This is very clear in Figure 8.2, which shows the genetic distance map for a number of Pacific Island, Asian, and New World populations. These distances were based on average distances of 120 genes of classic genetic markers of the blood and come from Cavalli-Sforza's global analyses, which were discussed in Chapter 5.[10] If there had been a recent historical connection between the New World and Polynesia, then we would expect to see genetic similarity between them. This is not the case. Instead, the genetic distance map clearly separates New World populations from those in Southeast Asia and the Pacific, with East

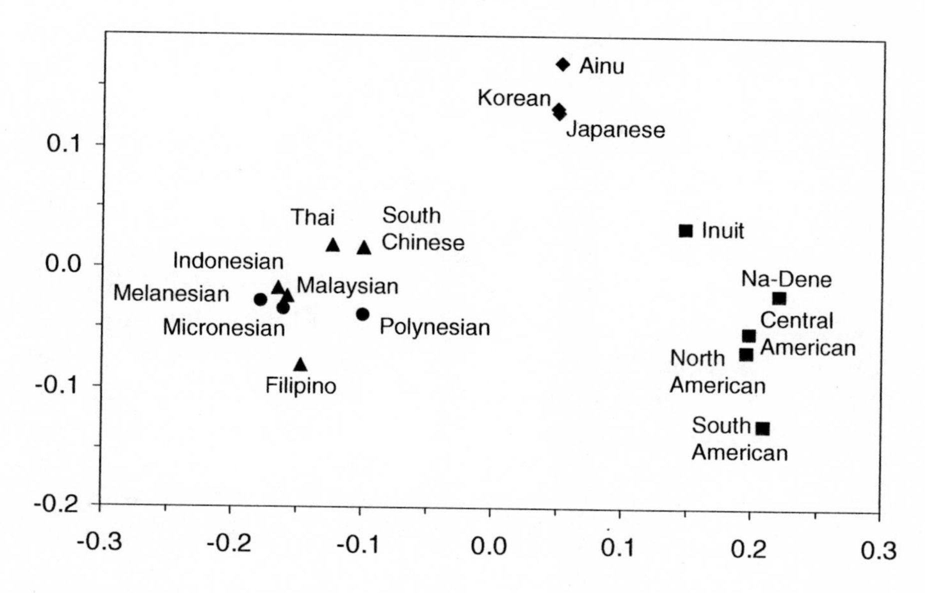

Figure 8.2 Genetic relationships between East Asian, Pacific Island, and Native American populations based on classic genetic markers of the blood. Triangles represent East Asian and Southeast Asian populations, squares represent Native American populations, and circles represent Pacific Island populations. This genetic distance map shows that Polynesia is quite distant from Native American populations, ruling out a New World origin of the Polynesians. Instead, Polynesians plot closest to Asian and other Pacific Island populations. The genetic distances are based on 120 alleles and are a subset of the distances that were presented in Chapter 5.
Source: Cavalli-Sforza et al. (1994).

Asian and South Asian populations in between. Polynesians are most similar to Asian and Pacific Island populations and quite different from Native Americans. Genetics does not provide any support for Heyerdahl's hypothesis of a New World origin of Polynesians.

Having ruled out a New World origin, what can we say about the relative merits of a model where Polynesians are descended exclusively from Asians versus a model incorporating gene flow from Melanesians? Are Polynesians more similar to Asian or Melanesian populations? If the express train model is correct, then we would expect to see close genetic relationships between Asians and Polynesians. To answer this question, I plotted the genetic distances between the Polynesian population and the other fifteen populations shown in Figure 8.2 to see which populations are genetically similar to Polynesians. Again, we see that the New World populations are the most distant from Polynesia (Figure 8.3). The Southern Chinese are the most genetically similar to Polynesians, which ties in nicely with the archaeological hypothesis of a Chinese or Taiwanese origin. However, the third most similar population is Melanesia, followed by the other Southeast Asian and East Asian populations. Does this mean that Polynesians have some Melanesian ancestry, or is this similarity due to Melanesians also having some genetic connection with Asia? Given only these overall genetic distances, it is not possible to distinguish between different models of Polynesian origins. These distances could fit a model of mixture between expanding Asians and Melanesians. However, they could also fit a model of exclusive Asian origin, with any resemblance to Melanesians being due to their sharing an earlier common ancestor in East Asia or Southeast Asia. In other words, if Melanesians came from Asia tens of thousands of years earlier, and then in more recent times the ancestors of Polynesians also came from Asia, then any similarity between Melanesians and Polynesians could be due to this remote common ancestry and not to a more recent mixture. The problem with the genetic distances shown in Figures 8.2 and 8.3 is that they could be explained by both models. We need another source of genetic information to resolve this question.

What Does Mitochondrial DNA Tell Us?

To answer these questions, we need genetic data that can tell us more about ancestral connections. As shown in previous chapters, a good way to look at such connections is through analysis of mitochondrial DNA. One approach is to look at a variant of mitochondrial DNA called the *9-bp deletion.* There is a specific section of noncoding mitochondrial DNA that lies between two genes. This section contains a sequence that is nine base pairs in length and looks like this: CCCCCTCTA. Many people have two adjacent copies giving them an 18-base-pair sequence of CCCCCTCTACCCCCTCTA. Other people lack one of these copies, a condition known as the 9-bp deletion because they are missing nine base pairs (bp). In these cases, one of their maternal ancestors had a mutation that resulted in the deletion of one of the copies, and they passed the deletion on to future generations through the female line.

The 9-bp deletion has been used a lot as a genetic marker for East Asian ancestry because it has been found in almost all East Asian populations as well as populations of Asian origin, such as Native Americans and Polyne-

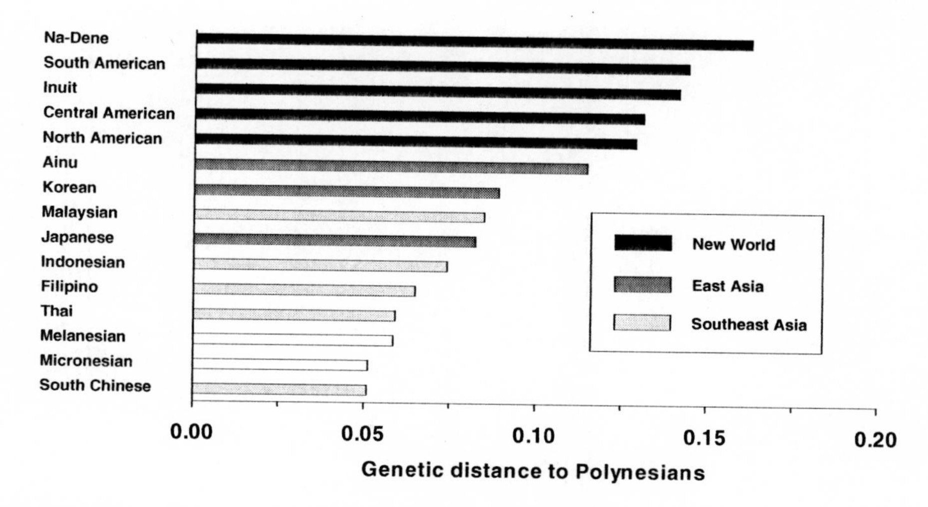

Figure 8.3 Who are most similar to living Polynesians? This graph shows the genetic distance to Polynesian populations for a number of East Asian, Pacific Island, and Native American populations based on classic genetic markers of the blood. Polynesians are most similar to populations in Southeast Asia and other Pacific Island populations (Melanesia, Micronesia) and perhaps represent a mixture of genes from both regions. This graph is based on the genetic distances in Figure 8.2.

sians. When we look at the spatial distribution of the 9-bp deletion in Polynesia, we see a geographic pattern that tracks the eastward movement out of Asia. The frequency of the 9-bp deletion increases from west to east along the route of Polynesian expansion (Figure 8.4). Although this pattern is consistent with the express train model, it doesn't rule out Melanesian gene flow, because some Melanesian populations also have the 9-bp deletion. Did the Melanesians acquire this trait during the expansion, or did they inherit it earlier from an Asian ancestor? If the former, then they gained this trait from gene flow out of the expanding Austronesian population. If true, then it seems likely that there was also gene flow in the opposite direction, from indigenous Melanesians into the population that was in the process of expanding into Polynesia. If the latter explanation is true, and Melanesians and Asians both inherited the 9-bp deletion from a common ancestor much earlier in time, then the presence in both does not necessarily imply gene flow from Melanesians during the Austronesian expansion.

The presence of the 9-bp deletion in many Asian and Pacific populations makes historical connections hard to resolve. To complicate matters, it now

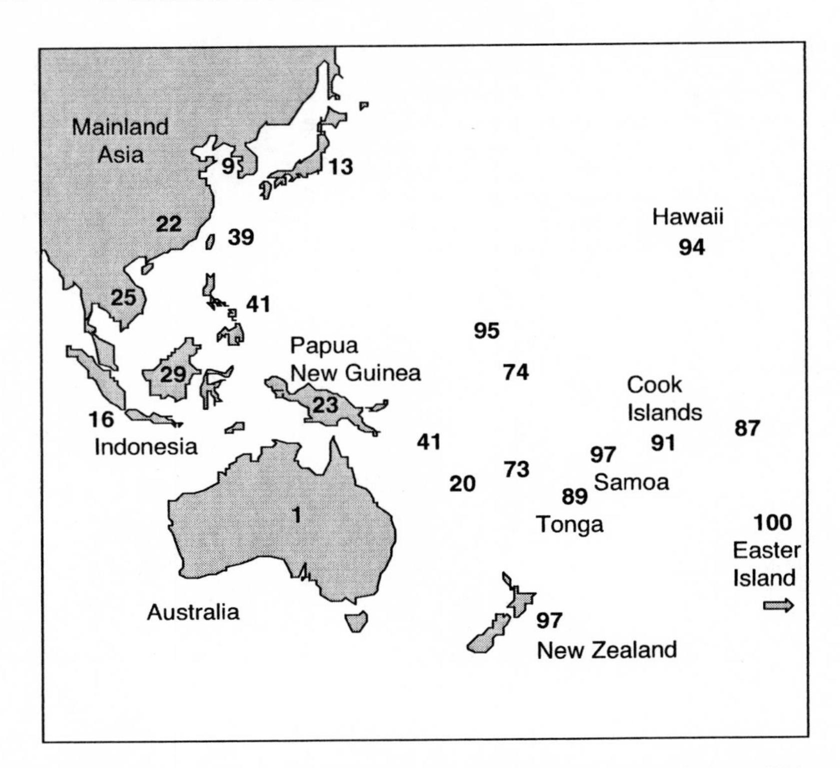

Figure 8.4 The 9-bp deletion and the origin of Polynesian populations. This map shows the geographic distribution of the 9-bp deletion in mitochondrial DNA in selected populations. The numbers refer to the percentage of individuals whose mtDNA had the 9-bp deletion. In general, the frequency of the 9-bp deletion increases from west to east along the route of the Austronesian expansion that colonized Polynesia. Deviations from this pattern most likely reflect genetic drift. Some population names are excluded to make the map easier to read.
Source: Merriwether et al. (1999).

appears that the 9-bp deletion is an example of a recurring mutation—a specific genetic change that happened more than once. You can see how this complicates historical reconstruction. If two people both have the 9-bp deletion, they might have both inherited it from the same maternal ancestor, or they might have inherited it from different maternal ancestors. The overall frequencies of this trait are not enough to settle the question of historical connection. To get a better idea of the specific historical connections, we need to look more closely at the mitochondrial DNA.

One way of doing this is by examining a variant of the 9-bp deletion common in Polynesians, where the 9-bp deletion occurs along with three specific sequence changes at other locations in the mitochondrial DNA, a pattern known as the *Polynesian motif*.[11] In one study, geneticist Alan Redd and his colleagues examined the geographic distribution of the Polynesian motif;[12] it has a low frequency in East and Southeast Asia but becomes more frequent moving east into the Pacific, reaching its highest frequency in the Polynesian population of Samoa. They argued that this spatial pattern is best explained by the express train model—an Asian origin and a rapid spread eastward. Another team of researchers, led by geneticist Bryan Sykes, came to a similar conclusion based on another mtDNA haplotype that also shared the 9-bp deletion.[13] Sykes and his colleagues looked at mitochondrial DNA in fourteen populations including Taiwan, the western Pacific, and Polynesia. The mitochondrial DNA haplotype they were looking at was also a Polynesian marker—it was found in 94 percent of Polynesians—and also showed the same spatial pattern of an eastward expansion out of Asia.

Although mitochondrial DNA suggests a strong affinity between Polynesians and Asians, as expected under the express train model, the situation is not quite that clear-cut. For one thing, while most Polynesian mitochondrial DNA sequences had the Polynesian motif, a small proportion did not, having instead sequences that suggested some Melanesian gene flow. Such gene flow is to be expected based on some earlier studies of hemoglobin genes (from nuclear DNA), which found mutations that were shared by Polynesians and Melanesians, something that would not be expected if the express train went by so fast as to preclude any genetic mixing.[14]

Could mitochondrial DNA be giving a different picture than other genes? If so, why? Geneticist Koji Lum and his colleagues looked at these questions by comparing genetic distances based on mitochondrial DNA with those based on nuclear DNA sequences (DNA sequences from the chromosomes in the nucleus).[15] Their study showed that the mitochondrial distances were similar to those found by other researchers—Polynesians were most similar to Asians—providing further confirmation of the express train model. However, their nuclear DNA distance analysis gave different results. Here, the Polynesian populations were genetically intermediate between Asian and Melanesian populations, suggesting genetic input from both sources.

What could account for these differences? One possibility is differences in mutation rates. The nuclear DNA markers used by Lum and his colleagues mutate faster than mitochondrial DNA, and this could conceivably give different pictures of population history. They ruled this out because of the short time depth; several thousand years is not long enough for variation in mutation rates to have had much effect. Another possibility they examined was genetic drift, which is greater for mitochondrial DNA than for nuclear DNA. The reason for this is simply the number of ancestors. With mitochondrial DNA, the number of individuals contributing to the next generation (mothers only) is half the number contributing with nuclear DNA (mothers and fathers). The smaller the numbers in the gene pool, the greater the potential effect of genetic drift. Deviations from historical relationships could happen by chance.

A third (and not mutually exclusive) possibility is that these two data sets were somehow picking up sex differences in migration. Remember that mitochondrial DNA picks up only the history of the maternal line, whereas nuclear DNA reflects the history of both parents. In other words, mitochondrial DNA picks up female gene flow, and nuclear DNA picks up both male and female gene flow. Lum and his colleagues suggested that their nuclear DNA data were showing the effects of male migration *after* initial Polynesian colonization. This could have happened because of genes being brought into the Polynesian population by *male* voyagers out of Melanesia. In other words, there was some genetic contact between male Melanesians and Polynesians *after* the express train roared past.

The Evidence from Y Chromosomes

The studies of mitochondrial DNA evidence have in general supported a predominately Asian origin of the Polynesians. Nuclear DNA, however, gives a different story, one that suggests greater Melanesian influence than shown by mitochondrial DNA. Lum and his coworkers' suggestion of sex differences in migration history is interesting, but analysis of nuclear DNA cannot get more specific because it picks up *both* maternal and paternal contributions. We need something to look specifically at the paternal contribution. As noted in previous chapters, the best way to do this is to examine genetic variations in the paternally inherited Y chromosome. Recent studies of Polynesian history have done this, and the results have shaken support for the express train model.

One such study was conducted by Manfred Kayser, of the Max Planck Institute for Evolutionary Anthropology, and his colleagues. They looked at genetic variants from the Y chromosome in the Polynesian population on the Cook Islands as well as seventeen other native populations in Asia, Melanesia, and Australia for comparison.[16] The first thing they found was that every single one of the twenty-eight Cook Islanders they studied had one of only three Y-chromosome haplotypes. The most common of these (known by the scientifically accurate but somewhat cumbersome name of DYS390.3del/RPS4Y711T) was found in 82 percent of the Cook Islanders. This haplotype is also in both Indonesian and Melanesian populations, although in lower frequencies (Figure 8.5). It is completely absent in East Asian, Southeast Asian, and Australian populations. The absence of this marker in East Asia and most of Southeast Asia runs counter to the express train model and argues for a strong Melanesian connection. If Polynesians were exclusively of Asian origin, then why isn't this haplotype found there, and why is it instead found in Melanesia?

Kayser and his coworkers feel that Polynesian origins are best explained by a "slow boat" model of Polynesian origins. According to this model, there was an expansion of Austronesian speakers out of Asia as postulated by the express train model. Unlike the express train model, however, the slow boat model proposes that movement through Melanesia was not so fast as to preclude genetic mixing of populations. It looks like the "train" was not an express; it made a few stops along the way. There was some gene flow from Melanesians into Polynesians, as well as the reverse. This mixing left some mitochondrial DNA variants, including the 9-bp deletion, in Melanesian populations while picking up Melanesian genes in the process. Other studies of the Y chromosome have arrived at similar conclusions.[17] The estimated dates of the common ancestors for these haplotypes are interesting because they suggest that some of this gene flow took place *during* the Austronesian expansion, and some occurred *after* the expansion.

Where in Asia?

Most everyone agrees that the Austronesian expansion started somewhere in eastern Asia, either on the eastern coast or farther south in mainland or insular Southeast Asia. Linguists have long pointed to Taiwan or the southern coast of China as a likely starting point. Most of the initial studies of mitochondrial DNA supported this idea, but this has recently changed. Martin Richards and his colleagues have challenged the traditional view, arguing instead that the specific point of origin was in Southeast Asia,

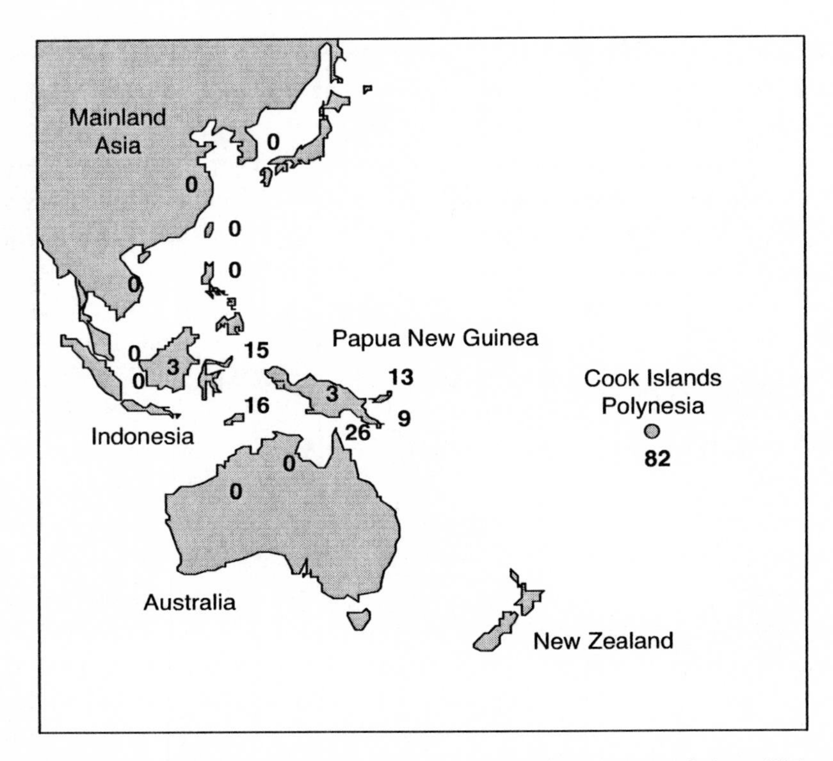

Figure 8.5 Y-chromosome DNA and the origin of Polynesian populations. This map shows the geographic distribution of the Y-chromosome haplotype DYS390.3del/RPS4Y711T in parts of Asia and the Pacific Islands. The numbers refer to the percentage of men having this haplotype. This particular haplotype is very common in the Cook Islander population in Polynesia (83 percent). It is also found in parts of Melanesia but is absent in East Asia, Southeast Asia, and Australia. The presence of this Polynesian marker in Melanesia supports the idea that some Polynesian ancestry comes from Melanesia.
Source: Kayser et al. (2000).

specifically farther south in the Indonesian archipelago.[18] They base their conclusion on their reconstruction of the evolutionary history of the Polynesian motif variant of mitochondrial DNA. Their analysis consisted of reconstructing a "family tree" of relationships between mtDNA haplotypes and then using the amount of diversity present in different populations to estimate the age of the most recent common female ancestor. They concluded that the Polynesian motif appeared first in eastern Indonesia roughly 17,000 years ago. It then spread to Polynesia as part of the expansion of Austronesian speakers.

Other recent analyses have also rejected the idea of a Taiwanese origin. For example, geneticist Bing Su and his colleagues examined Y-chromosome haplotypes in Taiwan, Southeast Asia, and Polynesia and found a very striking pattern.[19] The Southeast Asian sample (which included both coastal and island populations) showed the greatest range in haplotypes. The Taiwanese sample had *some* of the haplotypes found in Southeast Asia, but not all of them. In other words, the Taiwanese Y chromosomes were a subset of Southeast Asian Y chromosomes. The Polynesian Y chromosomes were also a subset of the Southeast Asian Y chromosomes, but a *different* subset from that in the Taiwanese samples. Thus, we have two populations (Taiwanese, Polynesians) that have different subsets of Y-chromosome variation found in Southeast Asians. This pattern is best explained by *both* Taiwanese and Polynesians having been descendants of a Southeast Asian population. Genetic drift leads to a subset of Y-chromosome haplotypes. When a small number of individuals leave a larger group, it is unlikely that they will possess the full range of genetic variation present in their "parental" population. If the Taiwanese and Polynesians were part of the same dispersion, they would both tend to have the same subset of diversity. Instead, Su and his coworkers found a different subset in each of these two populations, suggesting that these groups came from two different expansions out of Southeast Asia—one moving north to Taiwan and one moving east into Polynesia. This agrees with Richards's hypothesis of an Indonesian origin, since Indonesia was represented in the broad "Southeast Asian" sample collected by Su and his colleagues. It is also possible that the Southeast Asian connection to Polynesians could have originated elsewhere in this general region, including the southern coast of China, an area favored by some archaeologists as the birthplace of the Lapita culture.

Consensus?

Archaeologists have long argued that Polynesia was settled as part of the expansion of Austronesian-speaking farmers out of Asia, but that there has also been some mixing with Melanesian populations along the way. Overall, the genetic data show agreement with this model, although there remains debate over the relative contributions of Asian and Melanesian ancestry. The express train model, which suggests very little, if any, Melanesian ancestry, has to be rejected or at least modified substantially (a "slow boat" versus an "express train"). There is growing consensus today that both mitochondrial DNA and Y-chromosome DNA show evidence of mixed ancestry but in different proportions.[20]

Genetic analyses have suggested that Taiwan is unlikely to have been the source of Asian genes in Polynesians and point instead to Southeast Asia. Although some have suggested Indonesia as the specific source of these genes within Southeast Asia, I wonder about our ability to identify specific populations within a general geographic region. Any genetic analysis of *living* people must keep in mind that there have been changes in the genetic composition of those source populations over time. Can we use living Indonesians as a suitable proxy for Indonesians in earlier times, or has the genetic composition changed sufficiently in time, due to gene flow and genetic drift, to make this assumption somewhat less reliable?

Genetic studies of Polynesian history show an interesting sex difference in population relationships. Mitochondrial DNA shows a stronger signal of Asian ancestry, whereas Y-chromosome DNA shows a greater contribution from Melanesian populations. The fact that we get different results from mitochondrial DNA and Y chromosomes likely has something to do with sex differences in migration and mixture. If males and females in different populations mix in equal amounts, then we should see the same picture of genetic relationships in both mitochondrial DNA and Y-chromosome DNA. On the other hand, if there are definite sex differences in migration and mixture, then we are likely to see different patterns in these data. Studying mitochondrial and Y-chromosome DNA can help us unravel these gender specifics of population history. In the rest of our genes (in our nuclear DNA), we are a mix of maternal and paternal ancestry.

In the case of Polynesian origins, the data suggest that there was greater male than female gene flow from Melanesians. This might have happened during the initial expansion out of Southeast Asia past Melanesia. If there were more Melanesian men than women introducing genes into the expanding population, then there would be more impact on Y-chromosome DNA (inherited through the father) than on mitochondrial DNA (inherited through the mother). As a result, Polynesian mitochondrial DNA would be more similar to the original source in Southeast Asia, and Y-chromosome DNA would be more similar to Melanesia, which is the pattern we see today. Another possibility, favored by some, is that we are seeing the genetic effect of male gene flow out of Melanesia *after* the initial settlement of Polynesia.[21]

In other words, there was an expansion out of Southeast Asia past Melanesia (with some genetic mixture along the way) eastward into the Pacific. After this initial settlement, *male* voyagers from Melanesia made their way into Polynesia and mated with females, introducing Y chromosomes that are typically Melanesian. This scenario would also explain the differences between the mitochondrial and Y-chromosome analyses. It is also possible that *both* greater male gene flow during expansion and male gene flow after settlement occurred. I suspect that this may be the case. The history of any human population rarely reduces to a simple model of one-time mixture. Instead, we are likely seeing the joint imprint of a number of past migrations. As with examples presented in previous chapters, we are talking about changes among a set of populations that have probably been interconnected to some extent over time.

The potential complexity of Polynesian history is supported by genetic analyses of rats, of all things. As Polynesians spread, so did a number of animal species that they brought with them, including the Pacific rat *(Rattus exulans)*, most likely brought along intentionally as a potential food source. Skeletal remains of the Pacific rat have been found throughout Polynesia. Elizabeth Matisoo-Smith and her colleagues took mitochondrial DNA from living Pacific rats in order to get an idea of the rat's population history.[22] They found evidence of multiple introductions in different parts of Polynesia, including New Zealand and Hawaii, and argue that multiple contacts, rather than isolation following initial colonization, were the rule in Polynesian history. This evidence fits a growing body of evidence pointing to a complex pattern of colonization and postcolonization migration throughout the Pacific Islands.

A Thought about Expansions

Other than a brief analysis of the impact of geographic distance on allele frequencies on a Melanesian island (based on published data, not actual field research), I've never done any research on Pacific Island populations. Preparing for writing this chapter was largely the experience of an outsider looking in. However, my own research, which is primarily on the issue of modern human origins, has dealt with similar topics. I was struck by the close correspondence between the modern human origins debate (discussed in Chapters 3 and 4) and the question of Polynesian origins. In both cases, there was an acknowledgment of a genetic expansion, be it out of Africa or out of Southeast Asia. In both cases, there has been debate about whether there were other significant genetic sources other than from the expanding population.

A further parallel is seen in the previous chapter on the spread of farmers into Europe. Here too is a picture of a population expansion that did not wipe out previous inhabitants but instead mixed with them. Of course, there are many differences between all of these studies in both general and specific terms, but I was struck by the parallel concepts of expansion and mixture. Is this similarity simply a function of our models and methods biasing us toward a view of expansions, mixtures, and replacements, or does it speak to some common pattern? I suspect that both explanations are true to some extent, but I am most concerned with the commonality of process. The genetic history of human populations is largely the history of people moving and mating, so that most all populations today have a "mixed" ancestry.

In previous chapters, we have seen a number of examples of expansions during human history. In some cases, such as the initial spread of *Homo erectus* out of Africa or the initial movement of modern humans into the New World, we see evidence of expansion into an environment empty of humans. No one else was there. In others cases, such as in Europe during Neolithic times or the origin of Polynesians in the Pacific, we see a pattern of an expansion that mixes with the populations established during a previous expansion.

I suggest that the history of a species, particularly the human species, can be understood to some extent by examining the genetic impact of population expansions. On the other hand, I don't think that history is best explained by a series of independent populations going their separate ways until an expansion takes place. I instead suggest that there are two different processes going on throughout most of human evolution. First, populations rarely exist in isolation for a long time but are connected to each other through a web of interconnections maintained by gene flow. This gene flow varied from place to place and time to time, depending on a variety of factors, including physical and cultural barriers. Second, within this web of interconnected groups, there is sometimes a relatively rapid population expansion, perhaps due to some culturally derived advantage conducive to population growth, such as the development of agriculture. The expansion occurs along with genetic mixing, acting to reshape but not obliterate the previously established genetic landscape. Finally, the continued nature of local gene flow, shaped by isolation by distance, acts to further influence the distribution of genetic variants. Combine all of this with the action of genetic drift, most notable in small populations, and you get a mosaic pattern of genetic variation. None of this is new. Geneticist Alan Templeton uses similar concepts in describing the evolution of modern humans—recurrent gene flow constrained by isolation by distance and population expansions that result in interbreeding and not replacement.[23]

Much of human genetic history, including the Pacific Ocean populations, might be attributed to the joint influence of long-range dispersals acting together with local factors (migration, genetic drift). Although it is convenient at times, particularly in explanation, to simplify a complex set of historical forces, we should not forget the likely underlying complexity. In the case of Polynesian origins, much of the debate has centered on the relative contributions of Asian and Melanesian ancestry, expressed by models that start with an expansion out of Southeast Asia into the Pacific with some genetic mixture taking place along the way. Although this is a convenient summary of the genetic history of Polynesians, in some ways it is a bit simplistic because it assumes that two separate groups, Southeast Asians and Melanesians, had remained genetically isolated and homogeneous until some mixing occurred during the expansion of Austronesian farmers. In reality, there were probably varying degrees of interaction over time across the entire Pacific Island region. Archaeologist John Terrell has compared two different views of Pacific Island prehistory. One view he refers to as the notion of "history as family tree," where models deal with historical connections among separate and distinct populations. The other view, which he favors, is the idea of "history as an entangled bank," borrowing an ecological metaphor from Darwin's *On the Origin of Species* that focuses on the complex set of interactions likely to exist among populations.[24] Polynesian origins are best seen in an entangled bank of past history including a significant imprint due to the expansion of Austronesian farmers.

NINE

Three Tales from Ireland

"Would you like to go to Ireland this summer?"

I was asked this question in the winter of 1977 by my graduate advisor, Francis Lees. At the time, I was (literally) hanging out in the hallways at the State University of New York at Albany contemplating what to do next as a graduate student. I had reached that point in graduate school where I knew that I wanted to do research in anthropology and genetics but had no idea what to work on or how to start. Frank had been working on a number of projects with his former graduate advisor, Michael Crawford of the University of Kansas (my "academic grandfather"). One of these projects involved the reanalysis of data that had been collected by graduate students at Harvard University in the 1930s. Between 1934 and 1936, C. Wesley Dupertuis collected anthropometric data (measures of the body, head, and face) on more than 9,000 adult men throughout Ireland, and Helen Dawson collected many of the same measures on 1,800 adult females in the western part of Ireland.[1] Mike, who had already done some fieldwork in Ireland on a group known as the Irish Travellers (discussed later in this chapter), had been given access to the original data forms and was having them transferred to computer files for further analysis. Frank was working on this project with Mike and was also planning a research trip to Ireland.

In 1977, Frank was preparing for some preliminary fieldwork that summer in Ireland. One of the research goals of the project was the establishment of a long-term program for collecting anthropometric data, which would then be compared to the data from the 1930s. Frank was also interested in skin color variation, which anthropologists measured using one of two different reflectance spectrophotometers then available. Some anthropologists used one machine, while others used the second. These devices measured the amount of light reflected off an object at different wavelengths and proved an objective means of measuring skin color; the lighter the skin, the more light was reflected. Unfortunately, the readings of the two machines were incompatible, and this hindered efforts to determine the geographic distribution of skin color because the two sources of data could not be combined in the same analysis. To solve this problem, Frank had devised a conversion formula for dark-skinned populations, and now he wanted to extend it to light-skinned populations, such as those found in northwest Europe. Thus, a trip to Ireland would combine the beginnings of a new anthropometric study with the collection of skin color readings using both machines.

I knew little of this at the time. I was simply excited about the opportunity to do some fieldwork and was hoping that this would lead to a possible dissertation topic, although it was not at all clear how that would come about. Thus, I agreed to accompany Frank, along with Pam Byard, another graduate student at Albany. We spent the next few months learning how to take anthropometric and skin color measurements, and that summer we went to Ireland. We measured children at two schools in the city of Longford, County Longford, in the midlands of the Republic of Ireland. Frank wanted to start in Longford because the preliminary analysis of the Harvard data had suggested something unusual about this region of Ireland, a point I'll come back to later in this chapter. We focused on schoolchildren first, in part, to get a better idea of growth patterns among Irish children. A little over a month later, we returned to the United States with anthropometric and skin color data on more than 350 Irish schoolchildren.

My initial enthusiasm was quite intense. After all that reading and sitting in class, I was actually taking measurements on real people in another country. This enthusiasm was slightly dampened, however, after the first day of measuring children; such work is tedious at best. After the first dozen or so subjects, the thrill was gone. Nonetheless, I had a fine time, although upon my return some of my fellow graduate students, who had traveled that summer to much more isolated and far-off communities, commented that what I had done could hardly be called fieldwork. After all, we had plush beds, running water, and television!

By the end of our fieldwork, I was eagerly looking forward to the next step, which was analyzing the data. I have always loved data analysis, trying to make sense out of columns of numbers, seeking some pattern that would explain human variation. I could not do any of this in the field, as portable computers and other technological marvels that allow real-time data analysis did not yet exist. After we returned, I spent more than a week laboriously entering the data on computer punch cards (an archaic means of doing things, now thankfully behind us) and then several months analyzing the data. Eventually, these efforts led to several papers on anthropometric and skin color variation. None of these papers presented earth-shattering ideas, by any means, but their completion and publication did give me satisfaction.

This was all very well and good, but there was not enough substance for a doctoral dissertation, something I needed to graduate and get a job and leave behind endless meals of the peanut-butter sandwiches common to poverty-stricken graduate students. At the time, I thought that we would be returning to Ireland the following year, and I had begun to think of several possible dissertation topics focusing on the growth of Irish children. This was not to be, as Frank became involved in other projects and abandoned the fieldwork in Ireland. In the meantime, however, Frank was still working with Mike on the possibility of reanalyzing the 1930s Harvard data, and as Frank's research assistant, I got the task of collating and sorting these data. By this time, my interests were shifting away from human growth and back to my original interests in population genetics and human variation, particularly in two areas. First, I was interested in the general issue of using anthropometric data for studying population genetic models. Second, I was interested in the effects of geographic distance and cultural factors affecting migration on genetic variation among populations. I saw the opportunity to combine these interests through the analysis of the Harvard Irish data, and this became my dissertation topic.

My analyses dealt with data from twelve towns in the rural part of western Ireland and focused on the impact that geographic distance and migration had on the biological differences among these towns. This research focused on the genetic structure of these groups and did not really address genetics and history except for necessary background reading. Nonetheless, my interest in the human biology of Ireland had begun and led eventually to two different studies on the genetic history of Ireland.[2]

All of the above is a rather long prelude to the focus of this chapter—the genetic history of Ireland. The name "Ireland" refers in this chapter to the entire island of Ireland in the British Isles, which today is made up of two countries, the Republic of Ireland, which achieved independence in 1921, and Northern Ireland, still part of the United Kingdom. The Republic of Ireland is made up of twenty-six geopolitical units known as counties, and Northern Ireland contains six counties (Figure 9.1).

Figure 9.1 Map of Ireland. The island of Ireland consists today of two nations: the Republic of Ireland (twenty-six counties) and Northern Ireland (six counties). Adapted from a public domain map at www.irelandstory.com.

Archaeological evidence suggests that Ireland was first inhabited by hunter-gatherers about 9,000 years ago, followed by Neolithic farmers. A major influx of people began with the Celtic invasions roughly 2,500 years ago, which continued until about A.D. 300. Since that time, Ireland has seen a number of additional invasions and settlements, including invasion by Norse Vikings starting in the eighth century, followed by the Anglo-Norman invasions beginning in 1169. Between the fifteenth and seventeenth centuries, there was also a major influx of settlers from England and Scotland, primarily in the eastern part of Ireland.[3]

In addition to having experienced multiple invasions and settlements, Ireland is interesting because of the rapid changes in population size that took place during the nineteenth century. The demographic history of Ireland shows a classic "boom-bust" pattern of population increase and decline (Figure 9.2). The population began growing rapidly after 1700 due in large part to the introduction of the potato. Before this time, agriculture

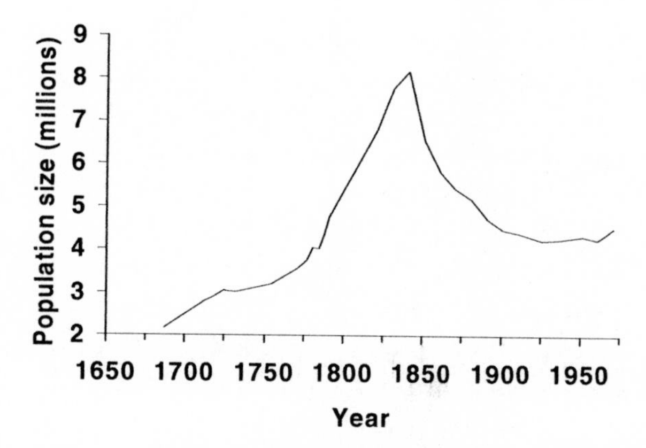

Figure 9.2 Changes in the population size of Ireland, 1687–1971. The population of Ireland increased dramatically after 1700 due to the introduction of the potato, which allowed greater agricultural productivity and, in turn, provided the opportunity for farmers to subdivide their land so that more than one son could inherit and thus be more eligible for marriage. Between 1846 and 1851 there was a major potato blight, leading to a rapid decline in population size due to death and emigration. Source: Vaughan and Fitzpatrick (1978). Data from the Republic of Ireland and Northern Ireland have been combined.

had been limited because Ireland has considerable hilly terrain and bogs that make some crops difficult to grow. The potato was a crop that could be grown in a wide range of conditions, and its introduction allowed a higher population size. The increase in population size was related to traditional patterns of marriage and inheritance in rural Ireland. Marriages were often arranged, and a usual prerequisite for marriage was for the groom to have land to farm and the bride to have a dowry. Because of the limited agricultural yield of the land, most families passed the land on to only one son; the other sons either remained unmarried or emigrated out of Ireland. When the potato was introduced, it provided greater agricultural efficiency, and it was possible for a father to subdivide the land and provide land, and thus an increased probability of marriage, to more than one son. Consequently, more sons married, and the population grew.[4] These changes also resulted in a precarious ecology that was very dependent on the successful continuation of the potato crop. Occasionally, the potato crop would be damaged by blight. The most serious consequence of this was the Great Famine (1846–1851), a five-year period of repeated infestation of the potato crop with no letup or time for recovery. During the Great Famine, an estimated 1.5 million people died, and an additional 1 million left Ireland.[5] Emigration continued in the following years, which kept the population from growing much during the twentieth century, even though fertility rates increased and mortality rates dropped.

The diverse history of invasion and settlement, combined with the more recent dramatic changes in population size, make Ireland an interesting place to look at the relationship between history and genetic variation. In this chapter, I discuss three examples of the use of genetic analyses in reconstructing the population history of Ireland. Two of these are based on my own work and look at historical patterns along the west coast of Ireland and across the entire island respectively. The first case study comes from research conducted by Mike Crawford and focuses on a historical question of the origin of a group of people known as the Irish Travellers.

The Origin of the Irish Travellers

Itinerant populations living a seminomadic lifestyle are found throughout Europe, ranging from the Woonwagonbewoners of Holland to the Taters of Norway to the many Gypsy populations found in eastern Europe. In Ireland, the itinerant population is known as the Irish Travellers, formerly referred to as the Irish Tinkers.[6] In 1996, there were almost 11,000 Travellers in the Republic of Ireland, constituting 0.3 percent of the total population.[7] In the past, they moved from town to town in horse-drawn wagons or by foot, repairing pots, pans, and other implements for money. Today, they live in encampments, mobile homes, and caravans and perform a variety of odd jobs and seasonal labor. The Irish Travellers live in small groups, usually consisting of several extended families. Overall, the Travellers are rather isolated culturally from the rest of the Irish population, and many speak their own language, known as Gammin.[8] They tend to have high fertility rates. Mike Crawford found that the average Traveller woman gave birth to more than ten children during her reproductive life, one of the highest levels of fertility recorded in a human population.[9]

There has long been historical debate over the origin of the Irish Travellers. Does the lifestyle of the Irish Travellers remind you of another group? To many, the itinerant lifestyle and social insularity of the Travellers suggest cultural similarity with European Gypsy populations. Could this cultural similarity be due to some actual historical connection? If they have a Gypsy lifestyle, then maybe they are Gypsies. Perhaps they were the descendants of Gypsies who moved to Ireland. This is, of course, a rather superficial comparison between two groups of people, but it might be taken as some evidence for a common origin.

Others have suggested that the Travellers are Irish. According to this view, the Irish Travellers are simply a subgroup within Irish society that resulted from farmers being displaced from their land by changing social and economic conditions. Although some historians have suggested this displacement took place following the Great Famine of the nineteenth century, most evidence suggests that the Travellers predate the famine and represent the long-term isolation of a segment of Irish society. Are the Irish Travellers Irish or Gypsy? The possibility has also been raised that they might be a little bit of both, that they have some Gypsy gene flow from elsewhere in Europe.[10]

When Mike Crawford conducted his research on the Irish Travellers in 1970, he realized that genetic data could help answer these questions. If the Travellers had experienced Gypsy gene flow, then they should be at least somewhat genetically similar to Gypsy populations in Europe and to the Punjabi peoples of northwest India, the area where Gypsy populations originated. On the other hand, if the Travellers were completely or predominately of Irish origin, then they should be genetically very similar to the rest of Ireland, allowing for some genetic drift due to social isolation and small population size.

To answer these questions, Crawford collected blood specimens from 127 Travellers in southeastern Ireland and an additional 95 specimens from unrelated Irish individuals living nearby. He looked at several classic genetic markers, including blood groups, serum proteins, and red blood cell proteins. Overall, the allele frequencies of the Travellers were very similar to those of other Irish and did not resemble those of Gypsy or Indian populations. A good example of his results is the frequency of the *B* allele of the ABO blood group. Gypsy populations typically have higher frequencies of *B* than other groups in Europe, owing to their origin in India, where the frequency of *B* is high. The frequency of *B* among the Travellers was 12 percent, identical to that from his unrelated Irish sample (12 percent) and similar to the results of an earlier nationwide survey of the Republic of Ireland (8 percent). Most Gypsy populations have higher frequencies of *B* (most range from 19 to 30 percent), as does the Punjabi population of India (28 percent). However, Crawford realized that *some* Gypsy populations did have lower frequencies of *B* and that relying on any one single gene could potentially be misleading. What he needed was an overall measure of genetic similarity, based on his full set of data, to compare Traveller, Irish, Gypsy, and Indian populations.

To get a better idea of the *average* genetic distance between the Irish Travellers and other populations, Crawford computed genetic distances between the Travellers and other European and Indian populations using data on thirteen different alleles. The genetic distance map is shown in Figure 9.3. Here, the Irish Travellers are most similar to other Irish and are not genetically similar to either Hungarian Gypsies or the Punjabi. The small difference between the Travellers and the rest of the Irish most likely reflects their social isolation and genetic drift. Hungarian Gypsies are closest to the Punjabi, which makes sense in terms of the Indian/Pakistani origin of Gypsy populations. The Travellers, on the other hand, show no specific similarity to Indians. Crawford therefore concluded that the Travellers are predominately of Irish origin.[11] Years later, two of Crawford's former students, Kari North and Lisa Martin, collaborated with Mike on a more extensive comparison of the genetics of the Travellers with those of other Irish. Again, the Irish Travellers showed clear-cut genetic similarity with the Irish samples, with no evidence of Gypsy ancestry.[12]

Crawford's study of the Irish Travellers is a good example of how genetics can be used to address the origin of an ethnic group. Moreover, I think it provides an excellent example of how culture and biology are often totally unrelated. In this case, the culture of the Irish Travellers suggested a link with Gypsy (and by extension Indian and Pakistani) populations. The similar seminomadic lifestyle and social isolation in this case are coincidental; they are two examples of a common process of social groups adapting to being dispossessed from mainstream society. Genetically, the Irish Travellers are not Gypsy but simply Irish. I was struck by the disparity between culture and genetics when I read Crawford's studies in graduate school, but I did not imagine that I would later find another example in my own research, also in Ireland.

English Gene Flow in the Aran Islands

After I finished my dissertation, I spent a year in a postdoctoral research position and then moved to my present position at Oneonta. During those years, I did no further research on Irish population genetics, as I was quite busy preparing to teach new courses and writing a series of articles based on my dissertation research. Tired of working on questions of Irish genetics, I switched my research focus quite abruptly and spent several

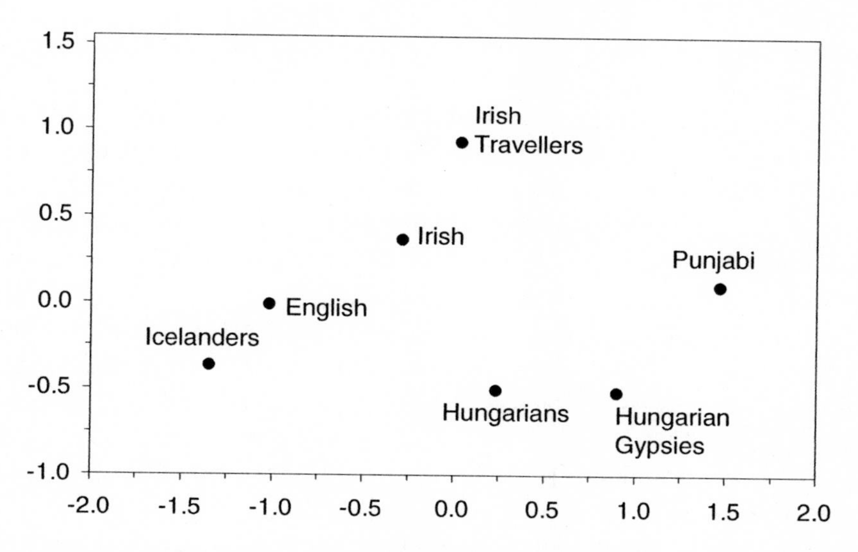

Figure 9.3 Genetic relationships of the Irish Travellers. This genetic distance map compares the Irish Travellers to other European populations. The Travellers are more similar to other Irish than to Hungarian Gypsies or to the Punjabi sample (representative of the place of origin of Gypsies). This analysis supports the hypothesis that the Travellers are of Irish origin and are not Gypsies.
Source: Crawford (1975).

years conducting historical demographic research of colonial America, using marriage records to investigate patterns of migration and inbreeding in several towns in north-central Massachusetts.

As the Massachusetts studies were winding down, I was reorganizing my files in my office and came across a number of research reports on the physical anthropology of Ireland that I had collected during background work for my dissertation. Included in these were a series of papers originally published in the 1890s in the *Proceedings of the Royal Irish Academy*, which dealt with anthropological investigations conducted in several islands and small villages on the west coast of Ireland. These papers included information on population size, folklore, customs, and just about everything else of interest to the investigators of Trinity College in Dublin. These papers also provided lists of surnames in each population. Because surnames are usually passed on through the paternal line, they provide a quasi-genetic marker of sorts, and I had used these lists in a preliminary study relating surname distributions to genetic distance in western Ireland.[13] Having rediscovered these papers in my files, I recalled an idea I had had while writing my dissertation, but which I did not pursue any further at that time. Lots of ideas come to you while writing a dissertation, but if you don't put them aside, you'll never finish.

The thing that had captured (and then recaptured) my interest was that the researchers in the 1890s had also collected anthropometric data, and they had published all of these raw data. At the time, I was interested in returning to my studies on Ireland and to some of the methods I had developed since my dissertation on estimating genetic distance from complex traits such as anthropometrics. To some, working on the relationship of anthropometrics to genetic distance might seem a step backward. After all, such measures are a step away from the underlying genetic code. Although measurements of the body, head, and face are a reflection of underlying genetic patterns, they are also influenced by aging, diet, and a variety of other nongenetic factors. Unlike DNA sequences or blood types, such measures change during a person's life.

Nevertheless, there had been a number of studies in anthropology that showed that such measures could be used under the appropriate conditions to estimate the underlying genetic differences between populations. Differences in age can be controlled for statistically. Comparisons of populations with similar cultural and physical environments can control to some extent for environmental variation. Selection of the right measures can also help; measures such as facial breadth are more likely to reflect populational differences than weight or body fat, which are strongly affected by diet and other factors. There was also a big advantage in using the data I had before me; anthropometric data was all they had back in the 1890s. Blood typing had not been discovered yet, let alone DNA. The data were right there, and all that I needed to do to look more closely at them was to enter all of the data into the computer. This required little time and no money.

When the task was completed, I had data on ten anthropometric measures of 259 adult Irish males who lived in seven different populations on or near the west coast of Ireland (Figure 9.4). Four of these populations are located close to each other in County Galway in or near Galway Bay. One of these populations, the Aran Islands, is located in the middle of the bay. Two small islands, Garumna and Lettermullen, lie close to the coast of County Galway, not far from the coastal village of Carna. The other three populations were located outside of the Galway Bay area. The island of Inishbofin is located a little bit north of this area, and two mainland villages, Ballycroy and Erris, are located much farther north in County Mayo. My initial expectation was that the genetic distance map would

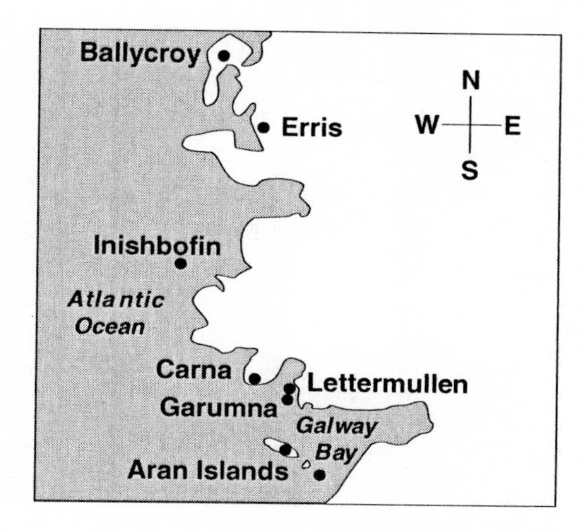

Figure 9.4 The west coast of Ireland. This map shows the location of seven west coast Irish populations for which anthropometric data were collected in the 1890s and which I used in one of my studies of Irish genetic history.
Adapted from Relethford (1988).

show a strong relationship with geography. After all, this is what I saw in my dissertation using data from farther inland. I expected to see three clusters: one consisting of the four populations in Galway Bay, which lie geographically close to each other, a second cluster consisting of the two populations in County Mayo, and the island of Inishbofin somewhere in between. Given the common finding of many studies of local genetic structure, this was a reasonable conclusion. Of course, this would not be very exciting, but it would give me a chance to compare parameters on a model I had developed for relating geographic and genetic distance to those from my dissertation data.

I did not exactly get the results I expected. Instead, the genetic distance map showed that the Aran Islands and Inishbofin were genetically distant from the other five populations (Figure 9.5).[14] If one ignored the Aran Islands and Inishbofin, one could see the expected geographic relationship, as the three populations in County Galway (Carna, Garumna, Lettermullen) clustered together, as did the two populations in County Mayo (Ballycroy, Erris). There was *some* of the pattern I expected to see, but the Aran Islands and Inishbofin clearly did not fit. I wondered what made them so different from the rest of the west coast populations, particularly their geographic neighbors in County Galway.

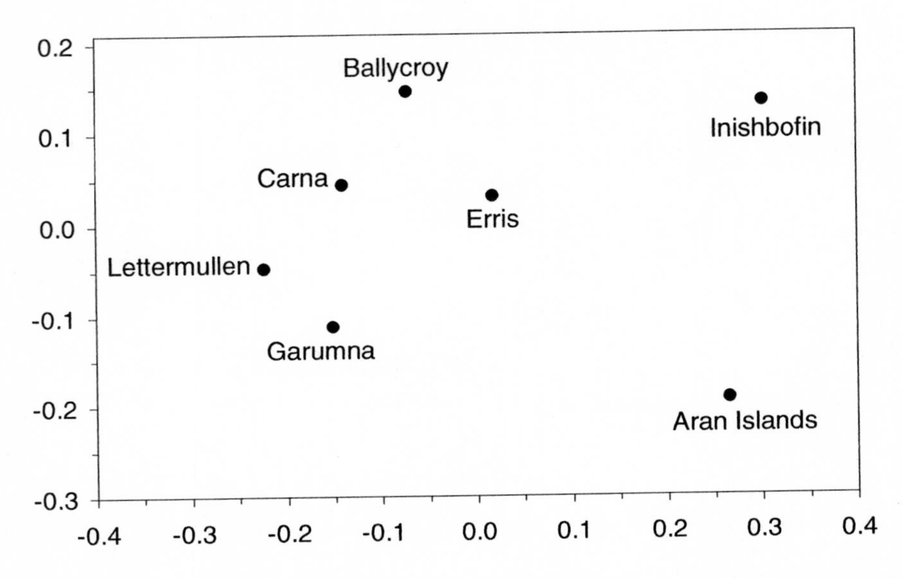

Figure 9.5 Genetic relationships in western Ireland. This genetic distance map of seven west coast Irish populations is based on ten anthropometric measures. The major pattern is the distinctiveness of the Aran Islands and Inishbofin. This distinctiveness is due to English gene flow from soldiers that were stationed on these islands over several hundred years.
Adapted from Relethford (2003) based on data in Relethford (1991).

I toyed with several ideas to explain this pattern, including the possibility that these two islands were more isolated due to their distance from the mainland, or perhaps they had experienced greater genetic drift in the past. However, none of these alternative explanations seemed to fit. I needed an explanation that would account for the fact that these two islands were genetically similar to each other, but different from the rest of the populations. In one of those interesting twists in research, I had actually read the answer earlier in my review of the history of the islands but had forgotten it. I decided to shelve the project for a time, coming back to it later on with a fresh mind. It turns out that *both* the Aran Islands and Inishbofin had experienced English gene flow over the course of several centuries.

In 1587, Queen Elizabeth I declared the Aran Islands part of her empire and granted them to John Rawson under the condition that he maintain English soldiers there. The first group of twenty soldiers arrived in 1588, starting an English military presence that was to last for several centuries, including some detachments from Cromwell's army in the seventeenth century. Historians have suggested that these islands had military significance as a possible defense of Galway Bay, guarding against pirates and military invasions from the Atlantic Ocean to the west. A similar set of circumstances occurred on the island of Inishbofin, where a garrison was stationed in 1652 that continued in existence for at least a century.[15]

The continued presence of English soldiers on these islands, and the likely interbreeding that would follow, suggests a likely route for the influx of English genes, which would explain the differences of the Aran Islands and Inishbofin from the remainder of the west coast Irish populations, who did not experience such gene flow. In other words, English gene flow made these populations dissimilar from the rest of western Ireland but similar to each other. This hypothesis was tested to a limited extent in 1958, when Earle Hackett, of Trinity College, Dublin, and M. E. Folan, of University College, Galway, collected blood samples from 229 residents of the Aran Islands.[16] They looked at the ABO and Rhesus blood groups in these residents and found that the distribution of blood types in the Aran Islands was in some ways different from that of the west coast of Ireland and were instead somewhat similar to those in England, suggesting the possibility of English gene flow. For example, 50 percent of the Aran Islanders had type O blood compared to 60 percent among the Irish living on the mainland of County Galway and between 42 and 53 percent in England. After looking at all different genes of both blood groups, Hackett and Folan found that the Aran Islanders were, on average, genetically intermediate between the English and the mainland Irish.

Hackett and Folan suggested that the genetic composition of the Aran Islands made sense in light of the popular belief in the west of Ireland, based on history, that "some of the ancestors of the Aran Islands (and also the people of Inishbofin) were men of Cromwell's garrisons."[17] They further concluded that the differences in the blood types reflected the fact that "There were military garrisons placed on Aran in the 16th and 17th centuries. The more permanent included English soldiers. They may have married island women and left descendants. The native population at the time of the military garrison was probably small, so any liaisons then could have made a notable genetic contribution to island stock which subsequently multiplied."[18] The Aran Islanders had some English ancestry because of these events.

The historical explanation of English gene flow introduced by soldiers over time fit my anthropometric analysis, which showed both separation of the Aran Islands and Inishbofin from the west coast of Ireland in the same direction, as expected if both populations had been acted on by the same evolutionary force—gene flow from England. What I needed to confirm this hypothesis was an anthropometric comparison with England. To do this, I compared my data with measures that had been collected by others on English populations. The results confirmed my hypothesis; the Aran Islands and Inishbofin were more similar to England than were the remaining Irish populations in my study.[19] These analyses, combined with the blood group study of Hackett and Folan, show that the Aran Islands and Inishbofin have mixed ancestry—some from English soldiers and some from Irish natives.

I was later able to further demonstrate the genetic effect of English gene flow by looking at the amount of anthropometric variation within my samples. Population genetics theory predicts that mixed populations will show greater diversity: the more outside gene flow, the higher the amount of genetic variation in a group. My colleague John Blangero and I developed a method for testing this theory using complex traits such as anthropometrics. This method allows one to find cases where there is more variation than expected, which typically means that a population has experienced more gene flow than average from outside the local area of study. We tried this method on my Irish data and found that both the samples from the Aran Islands and Inishbofin were more diverse than expected, which suggests they had experienced non-Irish gene flow. These results matched up with known history and my earlier analyses; the external gene flow was from England.[20]

I found a different pattern when I looked at the distribution of surnames among the seven populations. Surname frequencies can be used to estimate genetic distances and are analogous to measures based on Y chromosomes; that is, surnames are passed on only through the father's line. Since the source of English gene flow was from men, who pass on their surnames, I expected at first to see some evidence of the soldiers in the surname lists. When estimating genetic distances from surname frequencies, I found a moderate correspondence with geography, but no evidence of the English gene flow seen in anthropometrics and blood types. Why?

One possibility is that all of the English genes were passed on out of wedlock, in which case the father's surname would not be passed on as it would had couples married. More likely, the surname distances reflect the fact that surnames can disappear very quickly over short periods of time.

Imagine, for example, that you are the only person in a population with a given surname and all of your children are daughters, who take their husband's name upon marriage. Your surname would disappear. Over time, many relatively new or rare surnames can die out by chance. The rapid extinction of surnames seemed a reasonable explanation for what happened on the Aran Islands. I looked at surname lists recorded by Hackett and Folan and found that of the 135 surnames present on the Aran Islands in 1821, 67 percent had become extinct by 1892. I think that in general surnames change so rapidly, including name changes and extinction, that population history is wiped clean very quickly.[21]

As with Mike Crawford's Irish Traveller study, we see a bit of a disconnect between culture and genetics. The Travellers have often been seen culturally as Gypsies, yet they are definitively Irish in their genetic makeup. The Aran Islands and Inishbofin show the reverse pattern; these populations are culturally considered to be Irish yet they are less genetically similar to the rest of Ireland because of English gene flow. The west coast of Ireland, and in particular the Aran Islands, has often been regarded as having very traditional Irish culture. For example, the percentage of Gaelic speakers is highest in the western parts of Ireland. However, of all of Ireland, the populations of the Aran Islands and Inishbofin are the least similar genetically and represent a recent mixture of west coast Irish and English genes. The results of the long-term presence of the English soldiers had a noticeable genetic impact, but none culturally. The Travellers, on the other hand, show little change genetically from the rest of Ireland but have changed culturally. These examples show that changes in culture and genetics do not always coincide.

Invasions, Settlements, and Irish History

My work on the 1890s data reawakened my interest in the genetic history of Ireland and the use of anthropometric data for estimating genetic distances between populations. In my mind, I kept thinking about how interesting it would be to go back to the Harvard data that I had used for my dissertation, but to look at the entire island of Ireland and not just a small number of populations in the west coast region. The original analyses of the Harvard data, conducted in the 1950s, had suggested some interesting correlations with population history, notably the genetic impact of colonization, settlement, and invasion. I wondered if some of these hypotheses could be better answered today given advances in population genetic theory, statistical methods, and computer technology. When I wrote my dissertation, however, only a portion of the original data had actually been computerized. The time had come to get the rest of these potentially valuable data on line.

Although I felt a bit like a data scavenger, I wrote an application for research funds outlining the possible usefulness of these data, which had been collected many years earlier, and sent it to the National Science Foundation. In 1992, my grant was approved,[22] and I began the process of getting all of the original data coded, entered, and verified with the assistance of two of Mike Crawford's graduate students at the time, Ravi Duggarali and Kari North. By the end of that summer, they had retrieved and coded data for 8,385 males and 1,989 females. Because the adult female data were collected only in the west of Ireland, and because I wanted to look at population relationships across the entire island, all of my subsequent analyses were limited to the male data set. After eliminating individuals whose parents had not been born in Ireland, or who had missing data, I wound up with a sample of 7,228 adult males, each with complete information on seventeen anthropometric measurements of the body, head, and face. The next step was to remove the effects of age variation statistically, thus eliminating one potential source of variation that might confuse analysis and interpretation.

To look at variation between populations within Ireland, I used the county as the unit of analysis, which provided good sample sizes and retained much of the geographic variation across Ireland. I had to delete one county in the Republic of Ireland because of small sample size (County Wicklow), leaving me with estimated genetic distances between the remaining thirty-one counties. I looked at three sets of distance measures, one based on all seventeen anthropometric measures, one based only on the seven body measures, and one based only on the ten measurements of the head and face.[23] As I expected, based on results of other anthropometric studies, the analysis based on the head and face measures gave the clearest results. This is not surprising since body measurements such as height and weight often pick up other sources of variation due to diet, activity, and other nongenetic causes. All of the remaining analyses reported here are based on the head and face measurements.

One of the first things I looked at was a genetic distance map of the thirty-one counties (Figure 9.6). The first thing that is immediately obvi-

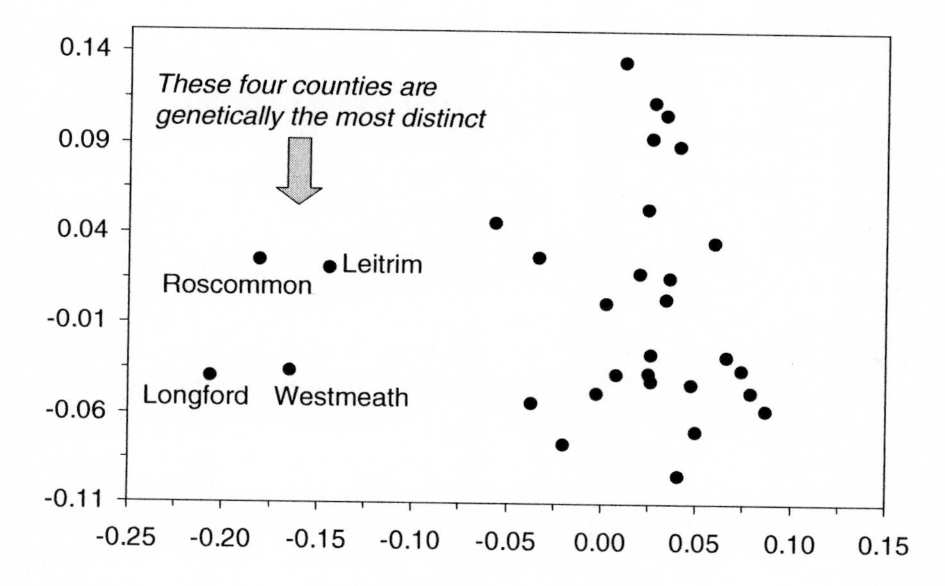

Figure 9.6 Genetic distinctiveness of the Irish midlands. This genetic distance map of thirty-one counties in Ireland is based on ten cranial and facial measurements. The most striking pattern is the genetic distinctiveness of four counties (Leitrim, Longford, Roscommon, and Westmeath), which are all located in the Irish midlands (see Figure 9.1). Adapted from Relethford and Crawford (1995).

ous from this distance map is that the major difference among the counties is the separation of four counties along the horizontal axis. They form a cluster that is distinct from the remaining twenty-seven Irish counties. These four counties (Leitrim, Longford, Roscommon, and Westmeath) are all located in the middle of Ireland. This pattern does not match up with geographic distance; if the geographic distance between populations was the primary cause of genetic differences, then we would expect populations in the center to be genetically central as well. Instead, they are the most divergent.

To be honest, I wasn't surprised. In fact, I rather expected to see something along these lines based on the preliminary reports from the Harvard group in the 1950s,[24] where they noticed a tendency for the midland populations to be different. However, analysis in those days was limited to looking at the geographic distribution of each trait one at a time, and the distinctiveness of the midlands was not always that clear. The distance map I generated used improved methods that looked at all traits at the same time, thus taking into account the correlation between traits. The result was a much clearer picture of midland differentiation.

The distinctiveness of the Irish midlands is most likely due to Viking invasions. The first Vikings came to Ireland in 794, landing on a small island near Dublin. Although many Viking settlements occurred along the coast of Ireland, their ships also sailed up the Shannon River from Galway Bay into the Irish midlands. In A.D. 832, the Viking warrior Tuirgeis led over 10,000 men to form the first permanent settlement of Norse Vikings. One of the major Viking headquarters was at a lake (Lough Ree) located in the midlands, although it was abandoned in 845. During the tenth and eleventh centuries, Ireland was again invaded by Vikings, this time Danish in origin. Again, many settlements took place along the coast, but some penetrated into the midlands along the Shannon River.

The movement of Norse and Danish Vikings into the Irish midlands could have had a substantial genetic impact on the midland populations as mating between Vikings and resident Irish took place. Genetically, this influx of genes would act to increase the genetic distance between the midland counties and the rest of Ireland, which is what is evident in the genetic distance map. One problem remained. *If* this is what happened, then why wasn't there a divergence of *other* parts of Ireland that also experienced Viking invasion, such as populations along the east coast? I suggest that in those cases the genetic impact of invasion was much less because later migrations into the coastal populations "erased" the previous traces of Viking gene flow. To investigate that possibility, I returned to the genetic distance map to see what other patterns of variation were apparent other than the distinctiveness of the midlands.

Looking more closely at the genetic distance map, I found that there was another pattern of variation related to longitude within Ireland: a noticeable gradient from west to east. Counties on the west coast of Ireland were somewhat distinct from those on the east. Counties in the western half of Ireland tend to plot together and to be farther away genetically from counties in the eastern half (Figure 9.7). In short, there is some genetic separation in Ireland between western and eastern counties.

A west-east division within Ireland has also been found in other genetic studies. The pattern shows up in classic genetic markers[25] and Y-chromosome haplotypes.[26] This geographic pattern is likely due to differences in the history of population settlement.

Some have suggested that the west-east distinction reflects the original settlement of ancient Celts, who may have entered Ireland in four waves of

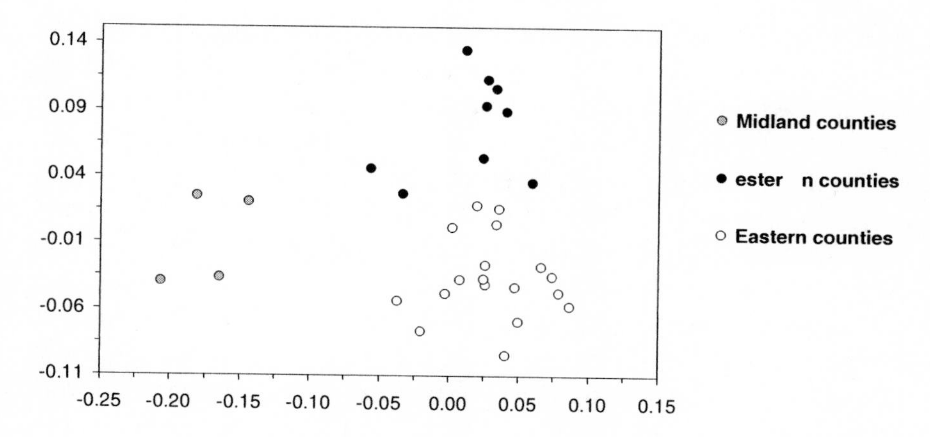

Figure 9.7 Genetic relationships in Ireland. This is the same genetic distance map as in Figure 9.6 but with the individual counties labeled to highlight geographic location. This genetic distance map shows two major features: (1) the genetic distinctiveness of the Irish midlands (gray circles), and (2) the clear separation between counties in the western half of Ireland (black circles) and in the eastern half of Ireland (open circles). Adapted from Relethford and Crawford (1995).

migration, with each new wave pushing earlier migrants farther to the west, thus causing a cline.[27] Others have suggested that the west-east cline within Ireland corresponds to the continued movement of farmers from the European mainland.[28] Settlements and population movements in more recent times have likely also contributed to this cline, such as the Anglo-Norman invasion of the twelfth century and settlement from England and Wales starting in the sixteenth century. The late Don Tills and his colleagues suggested that these later immigrants settled more frequently in the eastern parts of Ireland, with other immigrants from England and Scotland settling more frequently in the northern counties.[29] This kind of differential immigration, with people from different places settling in different areas of Ireland, could contribute to the west-east difference in genetic distances. In addition, these waves of immigrants could act to dilute the genetic effects of earlier population movements, perhaps explaining why there is no evidence of Viking influx in the coastal counties.

Overall, my anthropometric analyses pointed to two major historical influences on the genetic structure of Ireland: Viking invasions contributing to the distinctiveness of the midlands, and differential immigration and/or ancient population movements contributing to a west-east difference. To further test these ideas, I needed to compare the Irish data with anthropometric data from other parts of Europe, specifically Scandinavia and England. I found that the Irish midlands were the most similar of all Irish counties to both Norway and Denmark, the places where the Vikings came from. This observation confirmed the hypothesis that the distinctiveness of the midlands was in part due to the genetic impact of Viking invasion. I also found that the eastern counties of Ireland were more similar to England than the western counties were, in agreement with the suggestion that greater English immigration in the east had also had a genetic impact.[30]

The genetic history of Ireland provides a good example of two themes that I have touched upon in previous chapters. First, the genetic landscape of Ireland has not been shaped by any single event but instead reflects a mosaic of events that took place at different times, all overlaying one another to produce the patterns of genetic variation that we see today. Ireland was first settled by successive waves of newcomers from continental Europe, leading to the beginning of a west-east gradient in genetic differences. This gradient was likely enhanced by the arrival of the northwestern edge of the demic diffusion of agriculture through Europe (as described in Chapter 7). Viking invasions later left a significant impact on the Irish midlands, followed by the arrival of settlers from England and Wales, who settled predominately in the eastern part of the island. The genetic structure of Ireland reflects the joint impact of all these events, with particular events having greater relative impact in some areas, such as the impact of Viking invasion on the midlands. As such, the genetic structure of Ireland has changed over time. Different types of analyses can tease out these patterns, both ancient and recent, a common goal in all such studies of genetics and human history.

A second theme illustrated by these studies is the difference in focus depending on whether we look at things from a distance or up close. When we look at Ireland from a global perspective, it appears to be genetically typical of northwestern Europe. When we move in closer and focus on variation *within* Ireland, we see some major geographic differences, such as the distinctiveness of the midlands and the west-east difference. Looking even closer, we see interesting *local* variations that can also be explained in terms of history, such as English gene flow in the Aran Islands. Genetic changes are always taking place at local, regional, continental, and global levels. If we focused entirely on a global level of analysis (as in Chapter 5), we would see only the broadest picture and would miss the rich interplay of culture, history, and genetics that takes place within and between local populations.

TEN

Admixture, History, and Cultural Identity

The usefulness of the image of America as a "melting pot" continues to be discussed in academic circles. The term comes from the title of a play written in 1908 by Israel Zangwill, a Jewish emigrant from England. The metaphor is that immigrant cultures entering the United States will be forged into a new democratic culture, analogous to different metals combining into a new alloy when melted and mixed. The extent to which the United States functions as a melting pot continues to be debated by sociologists and others. Such discussions are part of a wider set of issues and questions concerning the dynamics of culture contact. What happens when different groups encounter each other?

There is a genetic component to culture contact. Population geneticists use the term *admixture* to refer to the mixing of genes from populations that hitherto had been separated for a long time. As we have seen in previous chapters, gene flow between human populations takes place over both short and long distances. Short-range gene flow takes place between neighboring populations, such as when mates are exchanged with a nearby village. Given sufficient time, short-range gene flow can spread genes across long distances, one step at a time (indeed, one name for this type of model is the "stepping-stone model"). Long-range gene flow can result from the expansion of populations, as seen in a number of previous chapters. In some cases, such as with the first Americans and the Polynesian settlement of the Pacific, this is simply the movement of people into an unoccupied region. In other cases, such as with the prehistoric spread of farmers in Europe, the migrants interbreed with the previous residents.

There are many examples of admixture in recent human history across the world. The past 500 years of human history have seen several examples of genetic admixture that resulted from the movement of Europeans from the Old World to the New World, beginning with Columbus's voyages of exploration and colonization at the end of the fifteenth century. Europeans moved into the New World and encountered Native Americans already living there. Enslaved Africans were brought to the New World during the slave trade. These historical events led to the genetic mixing of populations that had been separated in the past by time and distance. Whenever human groups encounter one another—whether peacefully or in confrontation—there is some genetic mixing.

Historical events leading to genetic admixture are not confined to the history of the New World. Such events occurred in the Old World as well. One example is the Jewish Diaspora, which took place more than 2,500 years ago. The geographic dispersal of Jewish populations led to varying degrees of admixture with their non-Jewish neighbors. This chapter examines several examples of genetic admixture in human populations.

The Genetics of Admixture

Nonbiological analogies of biological processes can be tricky but are often useful in understanding underlying principles. As in previous chapters, I use the mixing of paint as an analogy to the mixing of genes. Suppose you take two quarts of red paint and two quarts of white paint and mix them together. You wind up with four quarts of pink paint. If you wanted to get mathematical about it, you could describe the process of mixing as 50 percent red and 50 percent white because you used equal amounts of both colors of paints. What happens, however, if you take three quarts of red paint and only one quart of white paint? You will still get four quarts of pink paint, but the color will be a redder shade of pink because you used more red paint than white paint. The result can be described as 75 percent red (three out of four quarts) and 25 percent white (one out of four quarts).

We can imagine the same process with genes by dealing with the frequency of alleles in different populations. Imagine two populations, labeled "population A" and "population B," mixing together to produce a mixed (or hybrid) population, which we will call "population H." Let's further imagine that the frequency of a given allele is 0.9 in population A and 0.5 in population B. Now, let us assume that populations A and B each contribute equally (50 percent) to population H. What is the allele frequency in population H? Since both populations A and B contribute equally, the allele frequency in population H will be (½ × 0.9) + (½ × 0.5) = 0.7. In other words, the frequency in population H is midway between the frequency in A and the frequency in B.

What if the rates of mixture were different? For example, what happens if population A contributes 75 percent and population B contributes 25 percent? The allele frequency in population H will obviously be more similar to the frequency of population A than to the frequency of population B because population A contributed more of the genes. Specifically, the allele frequency in population H will be (¾ × 0.9) + (¼ × 0.5) = 0.8. Population geneticists generalize the process by using the symbol M to refer to the proportion of admixture from population B. Since proportions have to add up to 1, the proportion of admixture from population A is therefore 1 - M (Figure 10.1). In my first example, M = ½ and in my second example M = ¼. The higher the value of M, the more similar population H will be to population B.

These examples show how we can figure out the allele frequency in a hybrid population if we know the allele frequencies in the two "parental" populations and their rate of admixture. In studies of admixture and history, we use a similar method but work backward from observed allele frequencies. Given allele frequencies in parental and hybrid populations, we can estimate the amount of admixture that has taken place. For example, suppose we have a hybrid population with an allele frequency of 0.5, and we have historical evidence that the hybrid population was formed by the mixing of genes from two populations, A and B. Now, suppose further that we know that the allele frequency in parental population A is 0.4 and the allele frequency in parental population B is 0.6. What does this mean? Since the allele frequency in the hybrid (0.5) is exactly midway between the frequencies of A and B, we would conclude that A and B each contributed 50 percent to the gene pool of the hybrid. Using a simple formula, we can figure out the rate of admixture any time we have the allele frequencies of the two parental populations and the hybrid:

$$M = \frac{(\text{Frequency in A})-(\text{Frequency in H})}{(\text{Frequency in A})-(\text{Frequency in B})}$$

As another example, suppose we have allele frequencies of A = 0.8, B = 0.7, and H = 0.78. The proportion of genes from population B is M = (0.8-0.78)/(0.8-0.7) = 0.2. Therefore, the proportion of genes from population A is 1-M = 0.8. In other words, we have estimated that the hybrid population (H) is composed of 80 percent genes from A and 20 percent genes from B. To minimize sampling error, we would then want to do the same

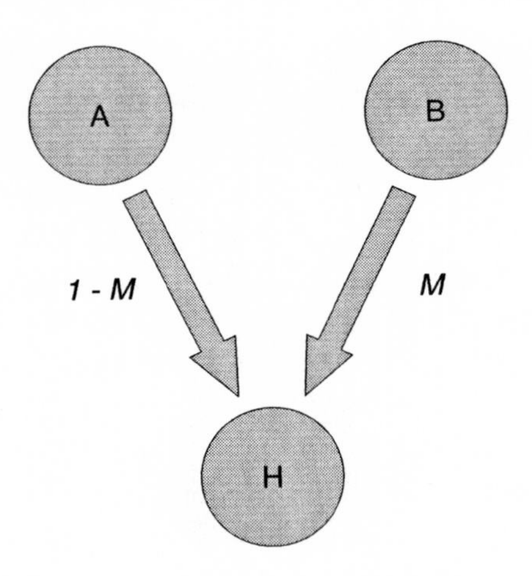

Figure 10.1 The process of genetic admixture. A hybrid population (H) is formed from the mixing of genes from "parental" populations A and B. The quantity M refers to the proportion of genes in H that came from B, and the quantity $1 - M$ refers to the proportion of genes in H that came from A. Since H is formed only from populations A and B, the proportions add up to 1 ($M + 1 - M = 1$).

thing on many other alleles and average their results. There are also more sophisticated methods that give statistically more reliable results, as well as methods that allow examination of more complex situations, such as a hybrid population made up of three, rather than two, parental gene pools.

It is important to keep in mind exactly what the terms M and $(1-M)$ represent. They are the *total* amount of accumulated ancestry from populations A and B in the hybrid population over some number of generations. These numbers by themselves say nothing about how long this process took. They could reflect a small amount of mixture in each generation accumulating over many generations, or they could reflect a larger amount of admixture occurring over a much shorter period. They could even reflect some combination of the two, perhaps a large amount to start with, followed by smaller amounts of admixture over a longer period. By themselves, these estimates provide only an estimate of how much admixture has taken place overall. Historical information would be needed to get a better idea of the time depth.

This method of admixture estimation holds a number of assumptions.[1] First, we have to be able to identify the ancestral populations. How did the hybrid population come to be? In some cases, historical data provide this answer. For example, we know from history that much of the population of Mexico reflects the mixing of European genes (mostly from Spain) with those of Native Americans. This knowledge allows us to use allele frequencies from living Spaniards and living (non-mixed) Native Americans to represent the genetic composition of the "parental" populations. However, in doing so, we come to the second assumption: Present-day allele frequencies accurately represent these populations in the past. In other words, we assume that there has been no significant evolutionary change in any of these populations over the given time period. This assumption is reasonably valid if the time depth is relatively short. Our third assumption is that admixture is the only factor that has affected the genetic composition of the hybrid population; that is, there has been no selection or genetic drift. This might be valid for some genetic traits but not for others, and we would have to apply appropriate methods to deal with this possibility. Finally, we assume that we have adequate sample sizes. Some of these assumptions hold up better in some studies than others.

Although somewhat simplistic, the admixture model does allow us to quantify the relative amounts of mixing between two (or more) populations that encounter each other and mix to produce a hybrid population. When combined with historical data, such studies give us a more precise analysis of evolutionary dynamics. Three examples of admixed populations are discussed in this chapter. Two of these resulted from historical events in the Americas over the past 500 years: the mestizo population of Mexico, and the African American population of the United States. The third example deals with the process of admixture between Jews and non-Jews following the Diaspora.

Admixture in Mexicans and Mexican Americans

Approximately 60 percent of the current population of Mexico is made up of mestizos, people who have "mixed" Spanish and Native American ancestry. The origin of the mestizo gene pool lies in the historical events following European exploration of the New World. In the early 1500s, Spanish explorers conquered much of the Native American civilization in Mexico, and waves of Spanish colonists followed. Considerable mating and intermarriage ensued over many centuries, resulting in the rapid growth of the mestizo population of Mexico, which became the largest ethnic group in the country by the beginning of the twentieth century.

Anthropologists and geneticists have conducted many studies on mestizo populations in Mexico in an effort to estimate admixture rates and understand population history. One such example is the work of Mike Crawford in the late 1960s and early 1970s on the Tlaxcaltecans, a group that lives in the Valley of Tlaxcala east of Mexico City. This area was conquered by the Spanish in 1519. Because of historical circumstances, some native populations remained relatively isolated, while those in the colonial administrative centers experienced admixture. Crawford began his research in two populations in the Valley of Tlaxcala: San Pablo del Monte, an Indian population that had experienced very little gene flow, and the city of Tlaxcala, which had experienced much more. He then became aware that there had been several relocations of Tlaxcaltecan populations into other parts of Mexico. One of these relocations consisted of a small number of Tlaxcaltecans sent to the town of Cuanalan, slightly north of Mexico City, to build a dike for Lake Texoco sometime prior to 1540. Another relocation took place in 1591, when ninety-one families were moved north to the Spanish colony of Saltillo. Crawford expanded his original research to include the populations of Cuanalan and Saltillo as part of a broad-based study of human biological adaptation that included focuses on genetic differentiation, impacts of different environments, and varying levels of admixture.[2]

One of Crawford's analyses estimated admixture in the sample from Tlaxcala. Most of the allele frequencies in Tlaxcala fell between those of San Pablo, representing a nonadmixed Indian population, and those of Spain (see Figure 10.2 for several examples). The intermediate allele frequencies in Tlaxcala would be expected if it were a hybrid population formed by Spanish admixture into the Native American gene pool. Furthermore, the Tlaxcala allele frequencies tended to be more similar to those of the San Pablo population than to those of the Spanish population, suggesting that the majority of the ancestry of this group was Native American (the genetically more similar parental population contributes more genes). Using methods that looked at allele frequencies over a number of genetic traits at the same time, Crawford and his colleagues estimated that the population of Tlaxcala was about 70 percent Native American and 30 percent Spanish in origin. Based on the history of contact, they further estimated that the 30 percent admixture rate worked out to an average of about 2 to 4 percent gene flow per generation from Spanish soldiers.[3]

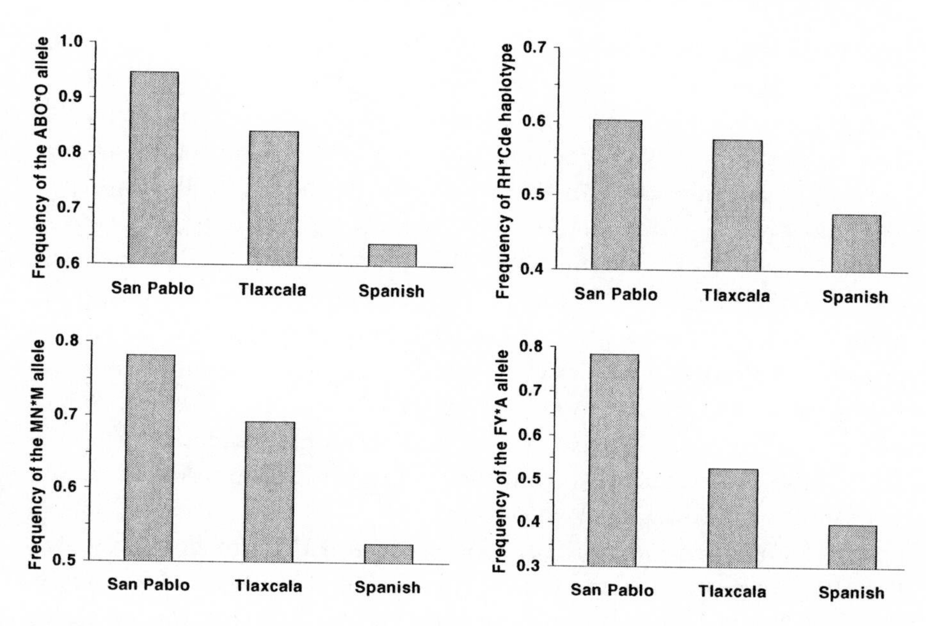

Figure 10.2 Admixture in Mexico. Frequencies of selected genetic markers in three populations: San Pablo, a Native American population; Spain, representing European genes brought into Mexico following conquest; and Tlaxcala, an admixed population that was formed by the influx of Spanish genes into a Native American gene pool. In each case, the allele frequencies for Tlaxcala lie between those of the Native American and Spanish "parental" populations.
Source: Crawford et al. (1976).

They also noticed an interesting pattern in Tlaxcala: a relatively elevated frequency of the Rhesus blood group haplotype called "cDe," which tends to be low in European and Native American populations, but which is much higher in African populations. The presence of this marker suggested that Tlaxcala had experienced some African gene flow. Based on this finding, and the fact that other studies of mestizo populations had also found some evidence of African gene flow, they looked at a model with three "parental" populations. They concluded that this model fit the genetic history of Tlaxcala better, suggesting ancestral contributions of roughly 70 percent Native American, 22 percent Spanish, and 8 percent West African. They suggested that the African admixture was due to Moorish ancestry among the first Spaniards or (more likely) introduced from enslaved Africans who lived throughout Mexico. A later study focusing on the Saltillo population in northern Mexico found a higher estimated proportion of African ancestry

(6–15 percent, depending on the specific sample), consistent with the historical fact that enslaved Africans were brought in to work at nearby mines.[4]

A large number of studies have shown that the mestizo population of Mexico was impacted by primarily Spanish admixture into the gene pool of Native Americans along with a small contribution from African genes. The rates vary somewhat from population to population, which is expected given local differences in the history of contact and cultural factors affecting gene flow. Studies of Mexican American populations in the United States also show a pattern of admixture, though with higher proportions of European admixture, likely due to continued gene flow from European Americans in the United States. Typical rates are about 50 to 60 percent European admixture,[5] although even here there is much variation. Admixed populations tend to be heterogeneous in terms of the relative rates of ancestry.

The comparison of classic genetic markers with mitochondrial DNA has allowed further insight into the dynamics of population admixture. Anthropologist Andrew Merriwether and his colleagues conducted such a comparison on Mexican Americans living in the San Luis Valley in Colorado.[6] Based on the analysis of classic genetic markers, they estimated rates of 67 percent European American and 33 percent Native American ancestry. The estimates from mitochondrial DNA, however, were quite different: 15 percent European American and 85 percent Native American. Because mitochondrial DNA is inherited only through the mother's line, the large difference in admixture rates suggests that the majority of mating in the past has been between European males and Native American females.

Admixture in African Americans

Small numbers of enslaved Africans were brought to the American colonies in 1619, and by 1700, importation of slaves was widespread. African slave importation increased throughout the eighteenth century but eventually diminished in the early nineteenth century after slave trading became illegal. It is estimated that between 380,000 and 570,000 Africans were enslaved and brought into the United States during this time. The overwhelming majority of enslaved Africans came from West Africa and West-Central Africa. Throughout American history, there has been genetic admixture into the African American gene pool, primarily as the result of

European men mating with enslaved African women.[7] Consequently, the African American population is an admixed population with at least two "parental" populations—West African and European—with, in some cases, additional admixture from Native Americans. Allele frequencies in the African American population tend to lie between those of West African populations and those of European populations, but closer to frequencies in West African groups, which is consistent with the majority of genes being of recent African origin and with a smaller proportion coming from Europeans.

Initial studies of European admixture suggested that roughly 20 percent of the gene pool of African Americans was European in origin. These early studies also showed that there was considerable variation in admixture proportions depending on several factors, including the choice of genes and choice of parental populations.[8] Some studies used historical information on country of origin, listed in shipping records of slaves, to weight the estimates of allele frequencies for the African parental population.[9] Even so, there was still variation in European ancestry among different samples of African American populations, showing that the concept of a single homogeneous African American gene pool was more a fiction than fact. Estimates of European admixture vary from 4 to 30 percent across African American communities.[10]

Recent studies have improved considerably on the accuracy and information that genetic data can provide about European admixture in African Americans. Esteban Parra and his colleagues have conducted several studies that rely on DNA markers known as *population specific alleles* (PSAs), which are either absent in one of the presumed parental populations or, if present in both, exhibit large differences between them.[11] Many genetic markers show extensive overlap among populations in different parts of the world and therefore are not very useful for differentiating relative ancestral contributions. The PSAs, however, show sufficient differentiation to allow more accurate reconstruction of population history.

Parra and his colleagues have collected data on ten nuclear DNA markers that qualify as PSAs for a number of African, European, African American, and European American populations. To illustrate the overall pattern of variation in these alleles, I used their published data to construct the genetic distance map shown in Figure 10.3. There is a clear separation between European (and European American) populations and African

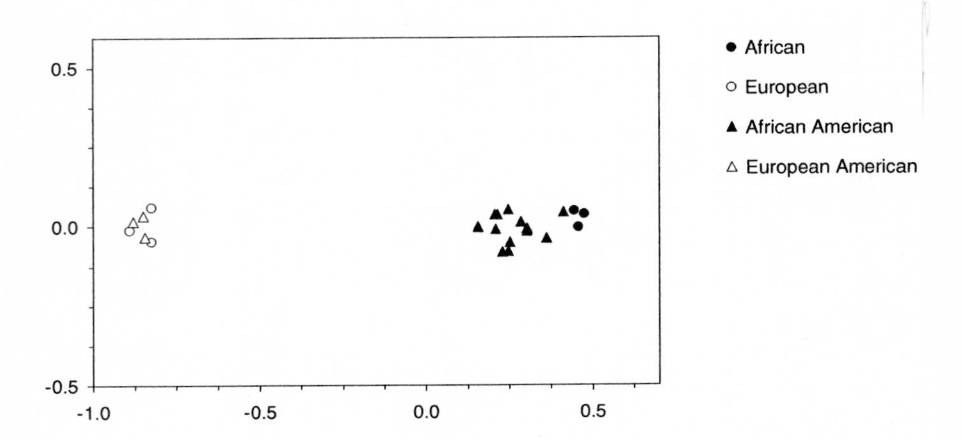

Figure 10.3 The genetic affinities of African Americans. Genetic distance map showing the relationships among a number of African, European, African American, and European American populations based on ten alleles that show high divergence between African and European populations. African American populations plot between European and African populations but are closer to the African populations, showing that the majority of these genes are African, but with European admixture. Genetic distances were computed using the Harpending-Jenkins (1973) method.
Source: Parra et al. (1998, 2001).

populations. The African American populations are somewhat separate from the African populations but much closer to them than to the European and European American populations. Again, this picture is consistent with a mixed ancestry of African Americans, but with a large contribution from West African genes and a small contribution from European genes.

Parra and his colleagues used the PSA data to estimate European admixture in twelve African American samples, ranging from rural to urban, and from northern and southern regions of the United States (see Figure 10.4). The average of these twelve samples is 16 percent European admixture, but there is a considerable range of variation in the individual estimates. The Gullah of South Carolina, a relatively isolated group of people living on small coastal islands, have the lowest (3.5 percent) amount of European admixture, and the African American community of New Orleans has the largest (22.5 percent). There seems to be a pattern of higher European admixture in urban areas. The three samples from South Carolina reveal a strong rural-urban difference.[12] The Gullah are the most culturally isolated and show by far the lowest

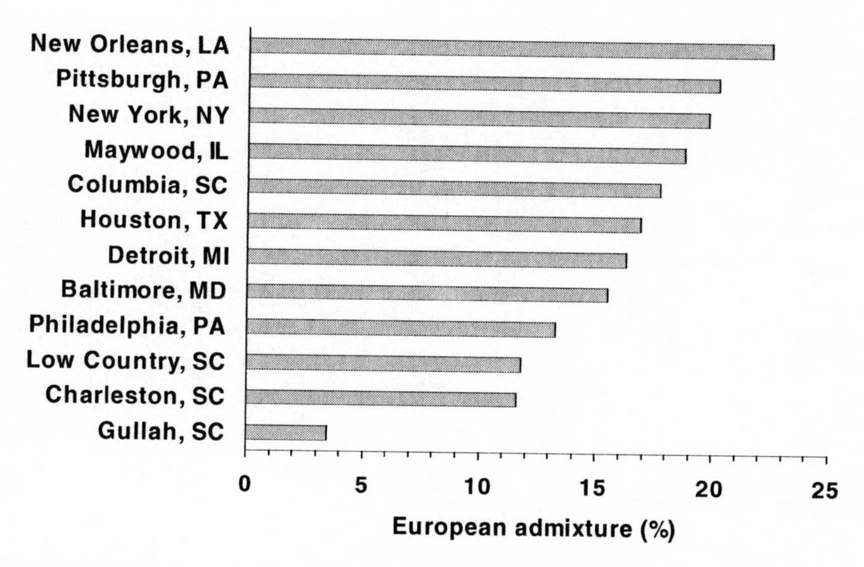

Figure 10.4 Estimates of European admixture in African American populations. There is a great deal of variation across samples in the amount of European ancestry, ranging from 3.5 to 22.5 percent.
Source: Parra et al. (1998, 2001).

amount of European admixture. The outer coastal plain of South Carolina, known as the Low Country, shows more European admixture (12 percent), and Columbia, a large urban area, shows the most (18 percent, a value similar to those found in northern cities). These data suggest that higher rates of European admixture are generally found in more urban areas, perhaps because of some urban-rural difference in attitudes regarding interethnic mating. More samples will be needed to be able to identify cultural, historical, and geographic factors that affect overall levels of European admixture (remember these estimates refer to the entire sample, and specific *individuals* within each sample could show considerable variation in European ancestry). For the moment, the existence of such variation across African American populations shows their heterogeneous nature. Any research study attempting to describe African American genetic variation in terms of a single number is problematic. Genetically, there is no single African American population, any more than there is any single European American population.

Parra and his colleagues also looked for sex-specific differences in admixture. That is, has European admixture in African American populations been primarily from females or from males? To answer this question, they used mitochondrial DNA to estimate the maternal component and Y-chromosome haplotypes to estimate the paternal component. As with the overall admixture estimates, there is a lot of variation across African American populations. What is striking, however, is that in each case, the amount of European admixture from the paternal line is greater than that for the maternal line (Figure 10.5). In other words, more European genes were introduced by men than by women into African American gene pools. This agrees with the history of enslaved African Americans, where male European slaveholders mated with enslaved African American women. Although there is some data suggesting that in recent times marriage between African American men and European American women is more common than the reverse, the history of enslavement was characterized by male European admixture.

It has also been suggested that there has been some admixture in the opposite direction, that is, the introduction of African genes into European American populations. Parra and his colleagues noted the presence of a typically African genetic marker, the Duffy Null allele, in three European American samples from Detroit, Pittsburgh, and Louisiana (Cajuns). Although the presence of this marker does suggest some African admixture, the actual estimates are low, averaging about 1 percent. It is clear that the history of culture contact in the United States has led to considerably more European American admixture in African Americans than the reverse.[13]

Finally, because earlier studies had suggested the possibility of Native American ancestry in some African American communities, Parra and his colleagues looked for the presence of the four typical Native American mitochondrial haplogroups (A, B, C, and D, as described in Chapter 6). In their first analysis of African American populations, they found these haplogroups in only 0.4 percent of the total sample, suggesting that there has been very little Native American admixture. They found somewhat higher proportions in their later study of South Carolina populations,[14] ranging from 1.1 percent in Columbia to 2.4 percent among the Gullah. However, analysis of Y-chromosome haplotypes in the South Carolina groups did not detect any evidence of Native American ancestry. Overall, it appears that Native American admixture has been low in most African American populations, and when it has occurred, it has tended to come from the maternal line.

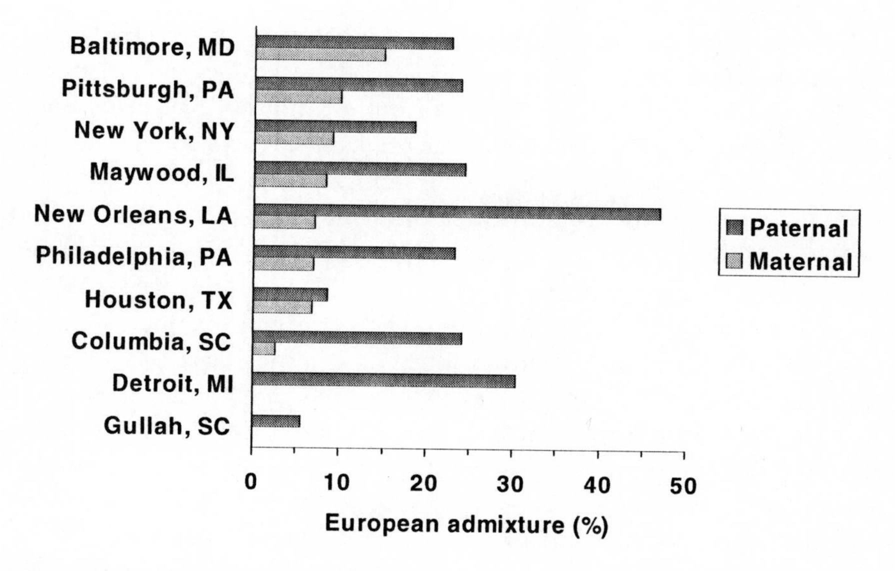

Figure 10.5 Sex differences in European ancestry of African Americans. This graph shows the estimated European admixture in the paternal line (based on Y-chromosome haplotypes) and the maternal line (based on mitochondrial DNA) in African American populations.
Source: Parra et al. (1998, 2001).

Thomas Jefferson and Sally Hemings

Perhaps the best-known recent controversy in admixture studies concerns the use of genetic data to investigate the paternity of the last child of Sally Hemings, an enslaved African American woman. Some historians have asserted that Thomas Jefferson, third president of the United States, was the father. In the mid-1780s, Jefferson's wife, Martha, died, and Jefferson was appointed ambassador to France. Sally Hemings, an enslaved African American, was sent to Paris to accompany Jefferson's youngest daughter, returning in 1789. During her life, Sally Hemings had at least five children. In 1802, Jefferson was accused of fathering Hemings's first child, Thomas, who was born in 1790 (Thomas was later named Thomas Woodson, after his subsequent owner). Jefferson is also suspected of having fathered Hemings's last child, Eston, who was born in 1808 and, it was said, showed a strong physical likeness to Jefferson. Although some historians have agreed with the suggestion that Jefferson was the father of Thomas and/or Eston, others have argued that the father was either Samuel Carr or Peter Carr, who were the sons of Jefferson's sister Martha.[15]

Was Thomas Jefferson the father of one or more of Sally Hemings's children? To date, the debate has focused on circumstantial evidence, including allegations of an affair while both were in Paris, the fact that Jefferson was in residence at Monticello at the time of conception, and Jefferson's physical similarity to Eston Hemings. Others have pointed out that other male members of the Jefferson family were also around at the presumed times of conception and might have been the father of one or more of Sally Hemings's children. If the father was related to Jefferson (as were his nephews Samuel and Peter), then that would explain the physical similarity between Thomas Jefferson and Eston Hemings. Thus, although the historical evidence can be argued to support the view that Thomas Jefferson was the father, it is not conclusive. These data support only the hypothesis that Jefferson *could* have been the father, not that he was indeed the father.

Can genetics provide a means by which to test for paternity? Yes, but the best methods are not applicable in this case, because everyone involved is dead. However, some tests can be made by examining the genes of their descendants (see Figure 10.6). This is exactly what Eugene Foster, a retired pathologist, and his colleagues did. They located male descendants of Jefferson, Hemings's sons Thomas and Eston, and the Carrs and compared Y-chromosome haplotypes based on a number of genetic markers. The object here was to compare genetic material passed through the male line to see how these descendants are related to each other. It was not possible to do this directly for Thomas Jefferson, because he had no known surviving sons. However, Foster and his colleagues were able to locate male descendants of Jefferson's paternal uncle, Field Jefferson. Because Field's father was Jefferson's paternal grandfather, they should have shared the same Y chromosome. The same applies to any further male descendants through Field's line unless there was a mutation somewhere along the way.

Four out of five male descendants of Field Jefferson shared the same Y-chromosome haplotype, and the fifth had a haplotype that was almost the same, except for one minor change that probably represents a mutation. The haplotype found in Field Jefferson's male descendants appears to be very specific to the Jefferson family and has not been found in large surveys of European men. The presence of a specific "Jefferson" haplo-

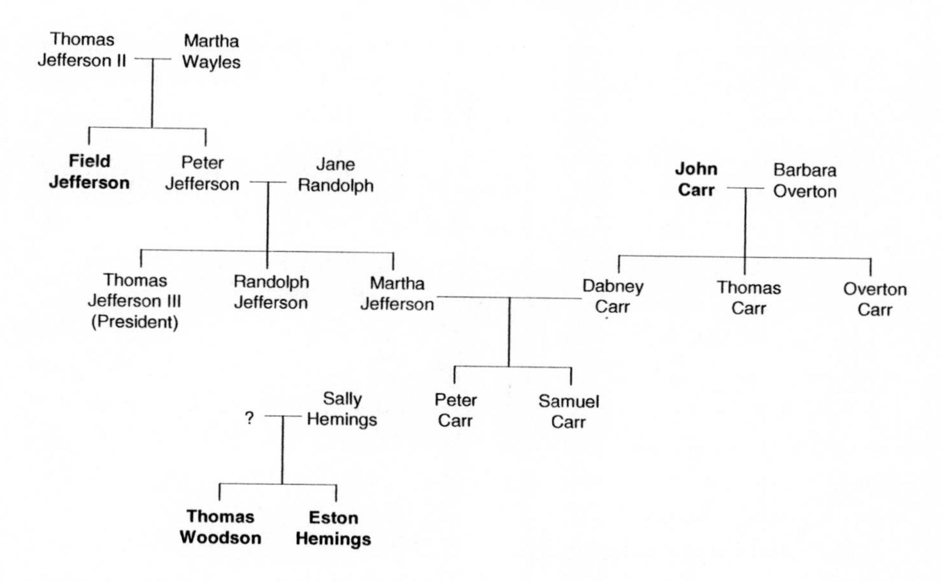

Figure 10.6 Who was the father of Sally Hemings's children? This graph shows a partial genealogy of the family of President Thomas Jefferson. Historians have suggested that President Thomas Jefferson, Peter Carr, or Samuel Carr was the father of either Thomas Woodson or Eston Hemings, both sons of Sally Hemings. The names in boldface show the individuals who have male descendants alive today from which Y-chromosome DNA was extracted.
Source: Foster et al. (1998) and Internet genealogies.

type provides a genetic marker that indicates paternal ancestry in the Jefferson family. The male descendants of Hemings's son Thomas Woodson do not have this haplotype, which is strong evidence that Thomas Jefferson was *not* the father of Thomas Woodson.

What about Eston Hemings? Foster and his colleagues were able to find only one male descendant of Eston, but this individual *did* have the Field Jefferson haplotype. On the other hand, none of the male descendants of John Carr, grandfather of Samuel and Peter, had this haplotype. Given that historical debate has focused on either Thomas Jefferson or one of the Carr sons as Eston's father, Foster and his colleagues concluded that the father was *not* Samuel Carr or Peter Carr and therefore was more likely to have been Thomas Jefferson. Although this was not the only possibility raised by Foster and his colleagues, it was certainly their favored conclusion, as indicated by the title of their article published in the November 5, 1998, issue of the journal *Nature:* "Jefferson Fathered Slave's Last Child."[16]

In reality, the use of Y-chromosome haplotypes can provide only a limited conclusion about paternity. In this case, it was sufficient to rule out the Carr brothers, because their male descendants had a completely different set of Y-chromosome haplotypes. Although the Carrs were related to Thomas Jefferson through his sister Martha, they had inherited their Y-chromosome from Martha's husband, Dabney Carr. Because most historical debate focused on *either* Jefferson *or* one of the Carr brothers as the father, and because the Y-chromosome analysis ruled out the Carrs, Foster and his colleagues concluded that the alternative hypothesis, that Jefferson was the father, was more likely. Although such paternity testing can help show someone is *not* the father, it does not necessarily show who the father actually was. The results of Foster and his colleagues show only that Thomas Jefferson *could* have been Eston Hemings's father. Could there be other possibilities?

Two letters arguing for alternative explanations appeared in a later issue of *Nature*. David Abbey pointed out that other potential fathers included Thomas Jefferson's brother Randolph and Randolph's five sons.[17] Any of the Jefferson males could have fathered Eston Hemings, because they all had the same Y-chromosome haplotype, inherited from Thomas and Randolph's father. In a second letter, Gary Davis suggested an even wider range of potential fathers.[18] He noted that *any* male ancestor in the Jefferson line, white or black, could have been the father. One possibility (among others) is that a paternal relative of Jefferson, perhaps his father or grandfather, fathered a male slave child whose descendant fathered Eston. The commonality of the Y-chromosome haplotype showed only that Eston Hemings's father was a male Jefferson or a descendant of a male Jefferson, not that he was a *specific* male Jefferson.

Foster and his colleagues responded to these letters, noting the validity of these criticisms, and added that some historical data supported the view that Isham Jefferson, one of Randolph's sons, may have been Eston's father. They concluded, however, that although the genetic evidence is not conclusive, the *historical* data favor the hypothesis that Thomas Jefferson was the father.[19] So, ultimately the case returns to consideration of the circumstantial historical evidence surrounding the case. Contrary to the title of Foster and his colleagues' initial article, the case has *not* been settled. The genetic data did help rule out the Carr brothers as possible fathers and showed conclusively that male descendants of Eston Hemings received their Y chromosome from *somewhere* in the Jefferson line.

Whether the specific source was Thomas Jefferson or a male relative remains unknown.

Genetic Admixture and the Jewish Diaspora

We now turn to an Old World example of genetic admixture by looking at the genetic relationships among Jewish and non-Jewish populations in the Middle East, Europe, and Africa. This example also touches upon the issue of cultural identity and its relationship to genetic affinity. Are Jews a biologically or culturally defined group? What are their genetic and cultural relationships with non-Jewish populations?

The origins of the Jewish religion and culture date back more than 4,000 years ago to the Middle East. The Diaspora began more than 2,500 years ago and marks the geographic dispersal of Jews to form communities outside of present-day Israel, primarily in Europe and North Africa. In 586 B.C.E. (an abbreviation for "before the common era," which is chronologically the same as B.C.), Jews were exiled to Babylonia by King Nebuchadnezzar II following the destruction of the first Temple in Israel. Since that time, Jews have been exiled repeatedly, forming communities in Europe, North Africa, and Southwest Asia. About 500 years ago, two major groups of Jews became somewhat isolated from each other: the Ashkenazim of central and eastern Europe and the Sephardim of the Iberian Peninsula and North Africa. The geographic dispersal of the Jews allowed the possibility for marriage or mating with non-Jews, and questions regarding the relative amount of genetic admixture have long been of interest to geneticists and anthropologists.

Much of the research in this area consists of comparing Jewish populations across the Old World with each other and with their non-Jewish neighbors. Do geographically separated Jewish communities tend to be more similar to each other or to their non-Jewish but geographically proximate neighbors? Greater similarity to each other implies a recent common origin and relative genetic and cultural isolation from non-Jewish populations since the Diaspora. If this is the case, then genetic distances between Jewish populations should be less than genetic distances between neighboring Jewish and non-Jewish populations. On the other hand, similarity to non-Jewish neighbors would imply genetic admixture, although such groups could retain cultural distinctiveness. In this case, genetic distances between geographic neighbors, both Jewish and non-Jewish, should be less than genetic distances between widely dispersed Jewish populations. Of course, one must also allow for the possibility that there has been *some* degree of isolation and *some* degree of admixture, perhaps varying from case to case, thus producing a set of genetic distances that is difficult to interpret.

Analysis of the genetic history of Jewish populations is further complicated for two reasons. First, we are looking at the genetic effects of events that have taken place since the Diaspora, which began only one hundred generations ago. Although this seems like a long time in terms of our own life span, in evolutionary terms this is the blink of an eye. How much genetic differentiation should we expect in such a limited amount of time? The second, and more difficult, problem to resolve is the fact that Judaism is a religion. Being Jewish is a condition of cultural membership that sometimes, but not always, is associated with a particular biological population originating among Semitic tribes more than 4,000 years ago. Examining the genetics of Jewish populations is difficult because Jewish identity is *culturally* defined. Not all Jews have recent genetic ancestry in the Middle East. I am a case in point; although Jewish, my own recent ancestry is probably more aligned with northwestern Europe than elsewhere. As far as I know, I have no recent ancestors who derived from the Semitic peoples of the Middle East. Nonetheless, I am still Jewish because I converted to Judaism, one of two major ways to be considered Jewish.

The other traditional criterion is through the mother's line; if your mother is Jewish, you are considered Jewish within the Jewish religion. This matrilineal connection has *some* connection with biology and genetic inheritance (except in the case of a woman converting). The children of a marriage between a Jewish mother and a non-Jewish father would be considered Jewish. The phrase "half-Jewish" has no meaning within Judaism. The important thing to remember here is that a population can remain culturally and religiously Jewish and still have two routes for genetic admixture: through conversion and through the father's line. There have been exceptions to the rule of matrilineal descent. Some, but not all, practitioners of the Reform branch of Judaism today allow patrilineal as well as matrilineal descent, although often with the proviso that the children be raised as Jews, something not required under traditional matrilineal descent. The point here is that group membership is often complicated and may not always have a direct relationship with concepts of genetic descent.

Returning to the basic question—how much non-Jewish admixture has taken place during the Diaspora?—what can genetic data tell us? Numerous studies have been undertaken using classic genetic markers to compare allele frequencies of Jewish and non-Jewish populations throughout Europe, North Africa, and the Middle East. Earlier studies have tended to show mixed results, with some analyses supporting rather extensive admixture from non-Jewish populations and others supporting the view of a common origin with relative isolation from non-Jewish neighbors.[20]

One of the most comprehensive genetic distance studies of Jewish and non-Jewish populations was conducted by geneticist Gregory Livshits and his colleagues Robert Sokal and Eugene Kobyliansky.[21] They compiled a large database of classic genetic markers for twelve Jewish populations in Europe, the Middle East, and North Africa. For each of these twelve populations, they also compiled data for geographically matched non-Jewish populations. They then compared the genetic distances between all populations with distances expected under different historical models, including common origin for Jewish populations, admixture with non-Jewish neighbors, and various combinations. They consistently found lower distances between pairs of Jewish populations than between pairs of non-Jewish populations, a pattern of genetic affinity that is best explained by a recent common origin of Jewish populations. However, they also found a secondary effect of non-Jewish admixture, with some tendency for Jewish populations to be more similar genetically to their non-Jewish neighbors than to non-Jewish populations farther apart. Overall, these data suggest that admixture has had some effect on Jewish populations but is not the primary factor affecting genetic diversity.

In recent years, there has been an increasing focus on DNA markers in studying Jewish population history. In one study, Michael Hammer and his colleagues looked at Y-chromosome haplotypes in seven Jewish and sixteen non-Jewish populations and found that the Jewish populations tended to cluster together with each other (Figure 10.7).[22] These Jewish populations also clustered with non-Jewish populations from the Middle East, such as Palestinians and Syrians, something found in previous studies and believed to be indicative of a shared origin among the Semitic tribes that inhabited the Middle East. The Ashkenazic Jews are the closest to the European populations, which would be expected if there had been some non-Jewish European admixture in the population. Hammer and his colleagues also looked at admixture in several of their samples and

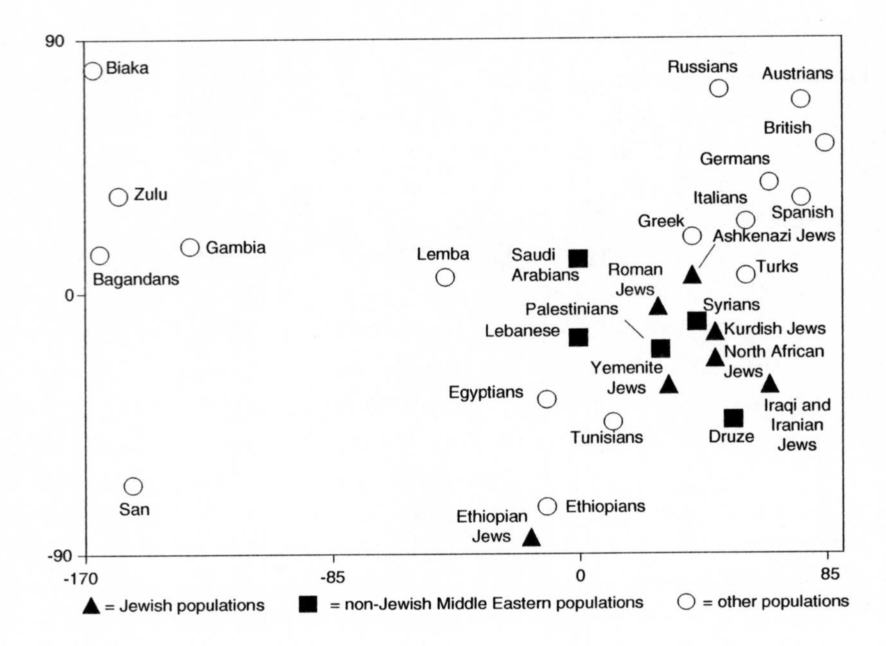

Figure 10.7 Genetic distance map of Jewish and non-Jewish populations based on Y-chromosome haplotypes. Most Jewish populations cluster together along with Middle Eastern non-Jewish populations. An exception is the Ethiopian Jews, who cluster closest to non-Jewish Ethiopians. The Lemba, a population in Southern Africa, is intermediate between African and Jewish populations. Adapted with permission from Hammer et al. (2000). Jewish and Middle Eastern non-Jewish populations share a common pool of Y-chromosome biallelic haplotypes. *Proceedings of the National Academy of Sciences, USA* 97:6769–6774.

estimated the amount of non-Jewish admixture at 13 to 23 percent for the Ashkenazim and at 20 to 29 percent for Roman Jews. Because these analyses are based on Y-chromosome haplotypes that are inherited through the male line, these are estimates of *paternal* admixture. Also, keep in mind that these are the *cumulative* admixture proportions over many generations; the amount of intermarriage within each generation was estimated to be less than 0.5 percent. Studies of mitochondrial DNA suggest less maternal admixture.[23]

Hammer's genetic distance map also shows interesting placements for two African populations. The Ethiopian Jews plot close to non-Jewish Ethiopians in Figure 10.7, distant from other Jewish populations. Similar results have been found when looking at mitochondrial DNA, which shows Ethiopian Jews to be distinct from Jewish populations in the Middle East and Europe.[24] These results are consistent with the view that the Ethiopian Jews converted to Judaism.

A second interesting population is the Lemba, a Bantu-speaking population of southern Africa often referred to as the "black Jews" of South Africa. The oral history of the Lemba suggests that they have paternal Jewish origins resulting from an influx of male Jews almost 2,000 years ago, who are thought to have moved into the area as traders and then married local women. The cultural evidence supporting their claim is largely circumstantial, consisting of some shared Jewish traditions. The problem here is that these traditions, which include certain food taboos and circumcision, are also found in other groups, including Muslim and African societies.[25] Can genetic evidence confirm the Lemba claims? Note that in the genetic distance map in Figure 10.7 the Lemba are genetically intermediate between sub-Saharan African populations, such as the Zulu, and Jewish populations in Europe and the Middle East. This position is consistent with a genetic link of the Lemba with Jewish populations. In another study of Y-chromosome haplotypes, Amanda Spurdle and Trefor Jenkins found genetic similarity between the Lemba and Semitic populations.[26] They estimated that roughly 40 percent of the paternal ancestry of the Lemba was of Semitic origin (which could mean Jews or Arabs).

An interesting analysis of Jewish history was conducted by Mark Thomas and his colleagues, who looked at Y-chromosome markers among two groups of male Jews: Levites and Kohanim (sometimes spelled Cohanim).[27] Jews consider themselves as belonging to one of three groups—Cohen, Levi, or Israel—depending on male ancestry. Most Jews belong to the category of Israel (including converts), but Levites and Kohanim (plural of Cohen) are believed to have descended from specific males and have special roles in religious observance. Levites are members of the tribe of Levi, who was a son of Jacob, and include Moses. In ancient times, the Levites served as priests but later were given other ceremonial functions when the role of priests was taken over by the Kohanim, who by tradition are the male descendants of Moses' brother Aaron. According to Jewish tradition, descent from Aaron has been passed from father to son over thousands of years and is often marked by the presence of certain surnames, such as Cohen and Kohen, among others.

The major question here is whether cultural identity correlates with genetic ancestry. Jewish tradition suggests a common ancestry of the Kohanim. If patrilineal descent occurred as suggested by tradition, from Aaron onward through male children, then all Kohanim share a common genetic male ancestor who lived several thousand years ago. If true, then analysis of Y-chromosome haplotypes should reflect this common ancestry. On the other hand, if the chain of descent had been broken or modified, and Kohanim were gained and lost over time, then there would be little if any correlation with genetic ancestry. To explore these issues, Thomas and his colleagues looked at the Y-chromosome haplotypes in 306 male Jews and asked them whether they were Cohen, Levi, or Israel. They then compared the distribution of haplotypes in these three groups.

They found one haplotype (named the Cohen modal haplotype) common in both Ashkenazic and Sephardic Jews who identified themselves as Kohanim. Fifty-four out of the 106 male Kohanim tested had this haplotype, which makes this sample very homogeneous genetically (Figure 10.8). Although the Cohen modal haplotype was also found among the Levites and Israelites, these other groups also had higher frequencies of other haplotypes, making them more diverse genetically. The genetic homogeneity of the Kohanim sample is consistent with common descent from a single male, as predicted from Jewish tradition. The presence of the Cohen modal haplotype in the Levites and Israelites could reflect common male ancestry among all Jews, predating the origin of the Kohanim and gene flow from Kohanim in the past. This does not mean that *all* Kohanim are direct male descendants of Aaron; the presence of other Y-chromosome haplotypes among the Kohanim suggests some male admixture, from Levites, Israelites, or non-Jews.

Thomas and his colleagues estimated the coalescence date, which suggests that the most recent common male ancestor of all Kohanim lived 106 generations ago. Assuming a generation length of twenty-five years, this places the origin of the Cohen modal haplotype at 2,650 years ago, a date that falls between the Exodus from Egypt and destruction of the first Temple in 586 B.C.E. They also noted that both Ashkenazi and Sephardic Kohanim showed high frequencies of the Cohen modal haplotype (45 and 56 percent respectively). This commonality, combined with the age estimates for a common ancestor, suggests that the Ashkenazim and Sephardim had common male ancestry before their initial isolation from each other 500 years ago, again consistent with known history.

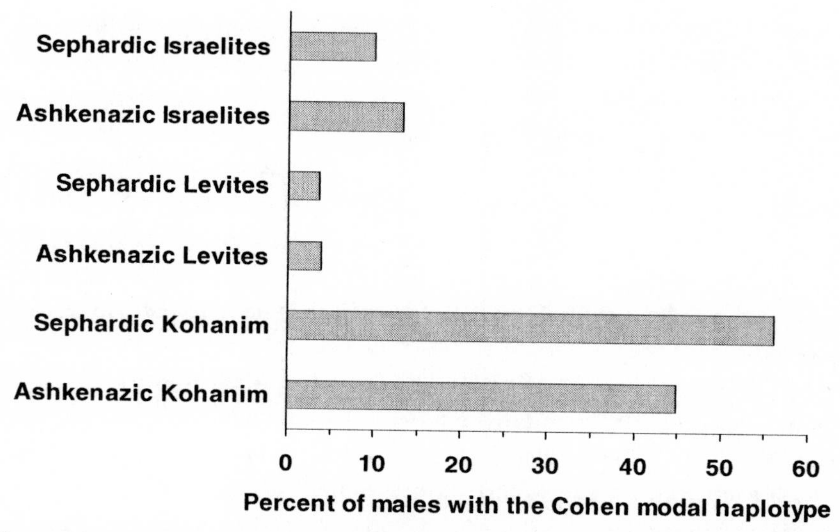

Figure 10.8 Frequency of the Cohen modal Y-chromosome haplotype in samples of Jewish men classified by origin (Ashkenazic versus Sephardic) and tribal membership (Cohen, Levi, Israel). The high frequency of this haplotype in Kohanim is evidence that these men share a common male ancestor, as expected based on Jewish tradition for the inheritance of tribal membership through the father's line.
Source: Thomas et al. (1998).

Analysis of the Cohen modal haplotype has recently shed some light on the issue of the paternal ancestry of the Lemba of southern Africa.[28] Mark Thomas found that roughly 9 percent of all Lemba men had this haplotype. The frequency was higher (over 50 percent) in Lemba men who belonged to the oldest clan, the Buba, the group identified through oral history as having led their ancestors out of Judea. The presence of the Cohen modal haplotype argues strongly for affinity with Jewish rather than Arab populations in the Middle East, as the marker is found in lower frequencies or is absent in Arab populations. However, they also found evidence of some male gene flow from non-Jewish Semitic populations. The Lemba appear to be an admixed group with genetic ancestry from both Jewish and non-Jewish Semitic populations as well as from African populations. On the other hand, the mitochondrial DNA of the Lemba does not show any Semitic admixture, which is also consistent with an oral history stressing *male* Jewish ancestors.

Are the Lemba Jewish? Although the Y-chromosome analyses show strong evidence of a genetic connection to Semitic populations that practice

Judaism, this is not the same as being Jewish. Jewish identity is most often defined culturally through conversion or maternal Jewish status. According to these rules, the Lemba are not Jewish culturally, even though they are related genetically to many Jewish populations through the paternal line. Jewish identity is a cultural condition, not a genetic or racial characteristic. Genetic studies show affinity between Jewish and non-Jewish populations in the Middle East, but this genetic affinity does not confer Jewish status on their Arab neighbors. Likewise, although there is a genetic and historical connection with the Lemba, this does not make them Jewish (unless they convert).

Overall, studies of the genetics of Jewish populations shows that the very nature of Jewish identity is only partially linked to a common genetic connection. Although Jewish populations have some genetic commonality with each other, they are also similar genetically to other Semitic peoples of the Middle East, a common origin expected from history. Here we have groups of people with different religions and cultures that are genetically very similar. In addition, not all Jewish populations share common genetic patterns, such as Ethiopian Jews. Jewish identity is culturally determined, resulting from maternal ancestry but also through conversion. Although there can be some biological connection through the mother (assuming she is not a convert), the paternal line can be from a completely different biological population and the children will be Jewish. Converts can come from any genetic background and be considered Jewish. The concept of a separate "Jewish race" in a biological sense is a myth. What we consider ourselves to be culturally does not always necessarily relate to genetic ancestry. Identity and ancestry can be thought of in both genetic and cultural terms, and the two will not always match up.

Genetic Ancestry and Cultural Identity

The chances are that you have many times been asked to fill out one form or another that asks for your "race." What box do you check off? A typical set of choices can be found on a census form, asking you to identify yourself as white, black, American Indian, Asian, Pacific Islander, or "other." In the U.S. Census 2000, you could also identify yourself as belonging to two or more races. In the course of filling out such forms, you might have asked yourself what the form was asking for—biology, culture, or both? The problem with the term "race" (and why it isn't used that much in studies of human population genetics) is that it has a number of different meanings, some of which conflict with each other. Sometimes the term is used in a biological sense, and sometimes in a cultural sense. Just think of the variety of adjectives that have been used with the term "race," such as "white race," "African race," "Italian race," or "Jewish race." "Race" mixes together classifications based on skin color, geography, nationality, and religion. Sometimes these different classifications might match up, but often they do not.

Anthropologist Jonathan Marks provides a good example of the confusion of genetic ancestry and cultural identity.[29] He describes reading a newspaper article written by attorney Lani Guinier in which she describes herself as being "black," whereas a photo caption in the article described her as "half-black." Marks wonders how someone can be both "black" and "half-black" at the same time. The answer is that the two statements refer to two separate things. The caption referring to Guinier as being "half-black" is a statement of recent genetic ancestry, in this case, referring to her parents, one of whom was white and the other of whom was black. Guinier describes herself as being "black," which is a statement of cultural identity reflecting her skin color, appearance, and the ancestry of *one* parent.

Studies of the genetics of African American populations show the very loose relationship between genetics and cultural identity. Although African Americans are often lumped into a single group based on cultural identity, their ancestry is variable, from place to place and from individual to individual. For example, Estaban Parra and his colleagues looked at genetic estimates of the ancestry of different African American individuals in Columbia, South Carolina. They found that although about half had less than 10 percent European ancestry, some had more than 50 percent.[30] Despite this genetic variability, all individuals classified themselves as African American. This shows the folly of equating cultural identity and ancestry. African Americans—or any group, for that matter—are not a homogeneous population. Nonetheless, the scientific literature is replete with studies of everything from IQ to disease incidence, which are then extrapolated to an entire group of people.

The studies on the genetics of Jewish populations show how cultural identity does not always map directly onto genetic ancestry. Being Jewish is a cultural condition; a person is Jewish if his mother is Jewish or if he converted to Judaism. A person whose mother is Jewish and whose father is not is considered Jewish in the cultural sense, and not "half-Jewish," a term that has no meaning in this context. Likewise, a person whose father is Jewish and whose mother is not would not normally be considered Jewish (except in some Reform congregations), again because identity is culturally determined. There is some connection with genetics because of the cultural transmission of Jewishness through the mother's line, but even here, it can be broken, as in the case of a woman who converts to Judaism. In some cases, the relationship between identity and genetics may be a bit clearer, as with the transmission of both identity and Y chromosomes in the Kohanim. In this case, identity is still culturally determined, but it can be tracked by a genetic characteristic. There is nothing about the Y chromosome that *makes* one Kohanim or Jewish; it is a coincidence. As with examples from previous chapters, such as the Irish ancestry of the Irish Travellers or the mixed ancestry of the Aran Islanders, cultural identity and genetic ancestry do not always go hand in hand. This is something we see more and more of in today's world, reflecting the separate natures of cultural and genetic change. Although cultural patterns can influence genetic variation, there is no evidence that genetic differences lead to cultural differences.

Genetics can tell us a lot about our biological ancestry and, combined with other data, can tell us about the history of human populations. I have attempted in this book to provide a number of examples of how anthropologists and geneticists have used genetic data to answer questions about population history. Again, these are our genetic roots, and they might be quite different from our sense of cultural identity or ethnicity or nationality. As a case in point, I would have to define myself in different terms depending on how the question, "What are you?" is asked. In terms of cultural identity and nationality, I'd answer, "American." In terms of *recent* ancestry I'd answer the same thing, since I have no ancestors that I am aware of who were born outside of the United States for several hundred years. Based on some limited family genealogical knowledge I can place some ancestors in England, although I don't feel particularly "English" or connected culturally to England in any way. In terms of ethnicity, I identify as Jewish, even though I am not aware of any recent biological connection to Semitic peoples. I suspect that many readers of this book have backgrounds that are equally mixed, both culturally and genetically.

One lesson we learn when studying genetics and human history is that our ancestry is more complex than was once thought. In earlier centuries, it was common to think of humanity as a set of races that had remained more or less separate and distinct since some distant time in the past. In fact, there was considerable debate over how to accommodate this view of human history with a biblical origin. Did all races trace back to a single one (and to a single set of individuals—Adam and Eve), or were there multiple creations of separate races? We now have a much richer and more complex understanding of the genetic history of the human species. The present is a reflection of the past, but not a simple one. Our genetic diversity reflects many events, some large and some small, both recent and distant in time. Our genetic history cannot be described as a simple family "tree" with parallel branches but instead is a tangled web of interconnections across time and space. As we look farther into the past, the pattern becomes even more mixed, and we all ultimately trace our ancestry back to Africa 2 million years ago (and perhaps more recently if the African replacement model is correct). In that sense, the study of genetics and human history provides us all, regardless of recent cultural or genetic background, with something in common. Our species' past belongs to all of us.

NOTES

Chapter 2
The Naked Ape

1. Kennedy (1976).
2. Goodman and Cronin (1982); Lewin (1997).
3. Goodman and Cronin (1982).
4. McKean (1983).
5. Sarich (1971).
6. Goodman and Cronin (1982).
7. Weiss and Mann (1990).
8. King and Wilson (1975).
9. Stoneking (1993).
10. Ruvolo et al. (1993); Horai et al. (1995); Gagneux et al. (1999); Chen and Li (2001); among others.
11. Morris (1967).
12. Sarich and Wilson (1967).
13. Sarich (1971).
14. Sarich and Wilson (1967); Sarich (1971).
15. Sarich (1971:76).
16. See Lewin (1997) for a lively discussion of this debate.
17. Horai et al. (1995); Gagneux et al. (1999); Chen and Li (2001).
18. Haile-Selassie (2001); Senut et al. (2001).
19. Gibbs et al. (2000).
20. Cela-Conde et al. (2000).
21. Morris (1967); Diamond (1992).

Chapter 3
Do You Know Where Your Ancestors Are?

1. Haile-Selassie (2001); Senut et al. (2001); Brunet et al. (2002).
2. See Wolpoff (1999) or any other current textbook on human evolution.
3. Swisher et al. (1994); Gabunia et al. (2000).
4. Wolpoff (1999).
5. Tattersall (1995), among others.
6. See, for example, the debate between Tattersall (1994) and Wolpoff (1994).
7. Relethford (2001c).
8. Wolpoff and Caspari (1997).
9. Stringer and McKie (1996).
10. Wolpoff (1999).
11. Lahr (1996).
12. Hawks et al. (2000); Wolpoff et al. (2001).
13. Cann et al. (1987).
14. Wainscoat (1987).
15. Penny et al. (1995), among others.
16. Templeton (1993, 1998).
17. Templeton (1993, 1997).
18. Hammer et al. (1998).
19. Ayala (1995); Harding et al. (1997); Harris and Hey (1999); Kaessmann et al. (1999); among others.
20. Templeton (2002).
21. Cann et al. (1987).
22. Relethford (2001c).
23. Hassan (1981).
24. Relethford and Harpending (1994).
25. Hassan (1981); Thorne et al. (1993).
26. Relethford and Jorde (1999).
27. Hassan (1981).
28. Harpending et al. (1993, 1998).
29. Whitlock and Barton (1997).
30. Relethford (2001c).
31. Relethford (2001c).

Chapter 4
The Fate of the Neandertals

1. King (1864).
2. See Trinkaus and Shipman (1992) and Stringer and Gamble (1993) for the history of Neandertal discoveries and interpretations.
3. Howell (1952).
4. Smith et al. (1999).
5. Duarte et al. (1999).
6. Tattersall and Schwartz (1999).
7. Stringer and Gamble (1993).
8. Bar-Yosef (1994).
9. Wolpoff (1999).
10. Krings et al. (1997).
11. Krings et al. (1999).
12. Ovchinnikov et al. (2000).
13. The paper by Ovchinnikov et al. reports 22 differences in the text, although Figure 2 shows 23 differences. The latter number is correct; the number "22" is a typographic error which should have read "23". William Goodman, personal communication, May 24, 2000.
14. Krings et al. (2000).
15. Krings et al. (1999).
16. Adcock et al. (2001).
17. Relethford (2001b).
18. Krings et al. (1997).
19. Krings et al. (1999).
20. Ovchinnikov et al. (2000).
21. Relethford (2001a).
22. Wolpoff (1999).

Chapter 5
The Palimpsest of the Past

1. Templeton (2002).
2. Hammer et al. (1998).
3. A general review of using physical traits in studies of population history can be found in Relethford (2003).
4. Major sources include Mourant et al. (1976), Roychoudhury and Nei (1988), and Cavalli-Sforza et al. (1994).
5. Cavalli-Sforza et al. (1994).
6. Wright (1969), among others.
7. Fix (1999).
8. Gagneux et al. (1999).
9. Barbujani et al. (1997); Relethford (1994).
10. Barbujani et al. (1997); Relethford (2002).
11. Relethford (2002).
12. Relethford (1997).

Chapter 6
The First Americans

1. Gould (1999).
2. Feder (2002).
3. Crawford (1998); Feder (2002).
4. Nemecek (2000).
5. Crawford (1998).
6. Bass (1995).
7. Cavalli-Sforza et al. (1994).
8. Schurr (2000).
9. Schurr (2000).
10. Derenko et al. (2001).
11. Schurr (2000).
12. Szathmary (1993); Merriwether et al. (1995); Schurr (2000), among others.
13. Powell and Neves (1999).
14. Greenberg et al. (1986).
15. Powell and Neves (1999).
16. Horai et al. (1993).
17. Merriwether et al. (1995).
18. Malhi et al. (2002).
19. Karafet et al. (1999).
20. Nemecek (2000).
21. Nemecek (2000); Marshall (2001).
22. Meltzer (1997).
23. Marshall (2001).
24. Nemecek (2000).
25. Thomas (2000).
26. Schafer (1996b).
27. Stang (1996b).
28. Schafer (1996a).
29. Stang (1996a).
30. Howells (1989).
31. Powell and Neves (1999).
32. Feder (2002).

Chapter 7
Prehistoric Europe: The Spread of Farming or the Spread of Farmers?

1. Population Reference Bureau (2001).
2. Feder and Park (1997); Price and Feinman (2001).
3. Ammerman and Cavalli-Sforza (1971); Ammerman and Cavalli-Sforza (1984).

4. Ammerman and Cavalli-Sforza (1984).
5. Menozzi et al. (1978); Cavalli-Sforza et al. (1993); Cavalli-Sforza et al. (1994); Piazza et al. (1995), among others.
6. Cavalli-Sforza et al. (1994); Cavalli-Sforza and Cavalli-Sforza (1995).
7. Richards et al. (1996).
8. Barbujani and Bertorelle (2001).
9. Chikhi et al. (2002).
10. Torroni et al. (1998).
11. Cavalli-Sforza et al. (1993); Cavalli-Sforza and Cavalli-Sforza (1995).
12. Piazza et al. (1995).
13. Sokal et al. (1992).
14. Cavalli-Sforza and Cavalli-Sforza (1995).
15. Sokal et al. (1996); Sokal et al. (1997).
16. Barbujani and Sokal (1990).

Chapter 8
Voyagers of the Pacific

1. Diamond (1999); Feder (2000).
2. Gibbons (2001).
3. Feder (2000).
4. Diamond (1999).
5. Bellwood (1991).
6. Diamond (1988).
7. Gibbons (2001).
8. Terrell (2001).
9. Sykes (2001).
10. Cavalli-Sforza et al. (1994).
11. Hagelberg and Clegg (1993).
12. Redd et al. (1995).
13. Sykes et al. (1995).
14. Kayser et al. (2000).
15. Lum et al. (1998).
16. Kayser et al. (2000).
17. Underhill et al. (2001); Capelli et al. (2001).
18. Richards et al. (1998).
19. Su et al. (2000).
20. Gibbons (2001).
21. Lum et al. (1998); Underhill et al. (2001).
22. Matisoo-Smith et al. (1998).
23. Templeton (1998); Templeton (2002).
24. Terrell (1988).

Chapter 9
Three Tales from Ireland

1. Hooton et al. (1955).
2. See Relethford (2003) for a detailed description of my interests and research in Ireland.
3. Hooton et al. (1955); North et al. (2000); Relethford (2003), among others.
4. Connell (1950).
5. Woodham-Smith (1962).
6. Gmelch (1977).
7. Web page of the Central Statistics Office, Ireland: http://www.cso.ie/publications/demog/travelcom.html
8. Gmelch (1977).
9. Crawford (1975).
10. Crawford (1975); North et al. (2000).
11. Crawford (1975).
12. North et al. (2000).
13. Relethford (1982).
14. Relethford (1988).
15. Browne (1893); Haddon and Browne (1893).
16. Hackett and Folan (1958).
17. Hackett and Folan (1958), p. 251.
18. Hackett and Folan (1958), p. 255.
19. Relethford (1988).
20. Relethford and Blangero (1990).
21. Relethford (1988).
22. The work described in this section was supported in part by Grant No. DBS–9120185 from the National Science Foundation.
23. Relethford and Crawford (1995).
24. Hooton et al. (1955).
25. Tills et al. (1977).
26. Hill et al. (2000).
27. Hooton et al. (1955).
28. Hill et al. (2000).
29. Tills et al. (1977).
30. Relethford and Crawford (1995).

Chapter 10
Admixture, History, and Cultural Identity

1. Reed (1969); Adams and Ward (1973).
2. Crawford (1976).

Chapter 10
Admixture, History, and Cultural Identity *(continued)*

3. Crawford et al. (1976).
4. Crawford et al. (1976, 1979).
5. Collins-Schramm et al. (2002).
6. Merriwether et al. (1997).
7. Chakraborty (1986); Parra et al. (1998).
8. Chakraborty (1986).
9. Adams and Ward (1973).
10. Chakraborty (1986).
11. Parra et al. (1998, 2001).
12. Parra et al. (2001).
13. Parra et al. (1998).
14. Parra et al. (2001).
15. Foster et al. (1998); Lander and Ellis (1998).
16. Foster et al. (1998).
17. Abbey (1999).
18. Davis (1999).
19. Foster et al. (1999).
20. Chakraborty (1986); Kobyliansky et al. (1982), among others.
21. Livshits et al. (1991).
22. Hammer et al. (2000).
23. Thomas et al. (2002).
24. Ritte et al. (1993).
25. Spurdle and Jenkins (1996); Thomas et al. (2000).
26. Spurdle and Jenkins (1996).
27. Thomas et al. (1998).
28. Thomas et al. (2000).
29. Marks (1994).
30. Parra et al. (2001).

REFERENCES

Abbey DM (1999) The Thomas Jefferson paternity case. *Nature* 397:32.

Adams J and Ward RH (1973) Admixture studies and the detection of selection. *Science* 180:1137–1143.

Adcock GJ, Dennis ES, Easteal S, Huttley GA, Jermiin LS, Peacock WJ and Thorne A (2001) Mitochondrial DNA sequences in ancient Australians: Implications for modern human origins. *Proceedings of the National Academy of Sciences, USA* 98:537–542.

Ammerman AJ and Cavalli-Sforza LL (1971) Measuring the rate of spread of early farming in Europe. *Man* 6:674–688.

Ammerman AJ and Cavalli-Sforza LL (1984) *The Neolithic Transition and the Genetics of Populations in Europe.* Princeton: Princeton University Press.

Ayala FJ (1995) The myth of Eve: Molecular biology and human origins. *Science* 270:1930–1936.

Barbujani G and Bertorelle G (2001) Genetics and the population history of Europe. *Proceedings of the National Academy of Sciences, USA* 98:22–25.

Barbujani G, Magagni A, Minch E and Cavalli-Sforza LL (1997) An apportionment of human DNA diversity. *Proceedings of the National Academy of Sciences, USA* 94:4516–4519.

Barbujani G and Sokal RR (1990) Zones of sharp genetic change in Europe are also linguistic boundaries. *Proceedings of the National Academy of Sciences, USA* 87:1816–1819.

Bar-Yosef O (1994) The contributions of Southwest Asia to the study of the origin of modern humans. In *Origins of Anatomically Modern Humans*, ed. by MH Nitecki and DV Nitecki. New York: Plenum Press, pp. 23–66.

Bass WM (1995) *Human Osteology: A Laboratory and Field Manual.* Fourth edition. Columbia, MO: Missouri Archaeological Society.

Bellwood P (1991) The Austronesian dispersal and the origin of languages. *Scientific American* 265(1):88–93.

Browne CR (1893) The ethnography of Inishbofin and Inishark, County Galway. *Proceedings of the Royal Irish Academy* 3:317–370.

Brunet M, Guy F, Pilbeam D, Taisso Mackaye H, Likius A, Ahounta D, Beauvilain A, Blondel C, Bocherens H, Boisserie J-R, De Bonis L, Coppens Y, Dejax J, Denys C, Duringer P, Eisenmann V, Fanone G, Fronty P, Geraads D, Lehmann T, Lihoreau F, Louchart A, Mahamat A, Merceron G, Mouchelin G, Otero O, Pelaez Campomanes P, Ponce De Leon M, Rage J-C, Sapanet M, Schuster M,

Sudre J, Tassy P, Valentin X, Vignaud P, Viriot L, Zazzo A and Zollikofer C (2002) A new hominid from the Upper Miocene of Chad, Central Africa. *Nature* 418:145–151.

Cann RL, Stoneking M and Wilson A (1987) Mitochondrial DNA and human evolution. *Nature* 325:31–36.

Capelli C, Wilson JF, Richards M, Stumpf MPH, Gratix F, Oppenheimer S, Underhill P, Pascali VL, Ko T-M and Goldstein DB (2001) A predominately indigenous paternal heritage for the Austronesian-speaking peoples of insular Southeast Asia and Oceania. *American Journal of Human Genetics* 68:432–443.

Cavalli-Sforza LL and Cavalli-Sforza F (1995) *The Great Human Diasporas: The History of Diversity and Evolution.* Reading, MA: Addison-Wesley.

Cavalli-Sforza LL, Menozzi P and Piazza A (1993) Demic expansions and human evolution. *Science* 259:639–646.

Cavalli-Sforza LL, Menozzi P and Piazza A (1994) *The History and Geography of Human Genes.* Princeton: Princeton University Press.

Cela-Conde CJ, Aguirre E, Ayala FJ, Tobias PV, Turbón D, Aiello LC, Collard M, Goodman M, Groves CP, Howell FC, Schwartz JH, Strait DS, Szalay F, Tattersall I, Wolpoff MH and Wood B (2000) Systematics of humankind. Palma 2000: A working group on systematics in human paleontology. *Ludus Vitalis* 12:127–130.

Chakraborty R (1986) Gene admixture in human populations: Models and predictions. *Yearbook of Physical Anthropology* 29:1–43.

Chen F-C and Li W-H (2001) Genomic divergences between humans and other hominoids and the effective population size of the common ancestor of humans and chimpanzees. *American Journal of Human Genetics* 68:444–456.

Chikhi L, Nichols RA, Barbujani G and Beaumont MA (2002) Y genetic data support the Neolithic demic diffusion model. *Proceedings of the National Academy of Sciences, USA* 99:11008–11013.

Collins-Schramm HE, Phillips CM, Operario DJ, Lee JS, Weber JL, Hanson RL, Knowler WC, Cooper R, Li H and Seldin MF (2002) Ethnic-difference markers for use in mapping by admixture linkage disequilibrium. *American Journal of Human Genetics* 70:737–750.

Connell KH (1950) *The Population of Ireland, 1750–1845.* Oxford: Clarendon Press.

Crawford MH (1975) Genetic affinities and origin of the Irish Tinkers. In *Biosocial Interrelations in Population* Adaptation, ed. by ES Watts, FE Johnston and GW Lasker. Chicago: Aldine, pp. 93–103.

Crawford MH (1976) Introduction: Problems and hypotheses. In *The Tlaxcaltecans: Prehistory, Demography, Morphology and Genetics*, ed. by MH Crawford. University of Kansas Publications in Anthropology, No. 7. Lawrence, KS: University of Kansas, pp. 1–5.

Crawford MH (1998) *The Origins of Native Americans: Evidence from Anthropological Genetics.* Cambridge: Cambridge University Press.

Crawford MH, Dykes DD, Skradski K and Polesky HF (1979) Gene flow and genetic microdifferentiation of a transplanted Tlaxcaltecan Indian population: Saltillo. *American Journal of Physical Anthropology* 50:401–412.

Crawford MH, Workman PL, McLean C and Lees FC (1976) Admixture estimates and selection in Tlaxcala. In *The Tlaxcaltecans: Prehistory, Demography, Morphology and Genetics*, ed. by MH Crawford. University of Kansas Publications in Anthropology, No. 7. Lawrence, KS: University of Kansas, pp. 161–168.

Davis G (1999) The Thomas Jefferson paternity case. *Nature* 397:32.

Derenko MV, Grzybowski T, Malyarchuk BA, Czarny J, Miścicka-Śliwka D and Zakharov IA (2001) The presence of mitochondrial haplogroup X in Altaians from south Siberia. *American Journal of Human Genetics* 69:237–241.

Diamond J (1988) Express train to Polynesia. *Nature* 336:307–308.

Diamond J (1992) *The Third Chimpanzee: The Evolution and Future of the Human Animal.* New York: HarperCollins.

Diamond J (1999) *Guns, Germs, and Steel: The Fate of Human Societies.* New York: W. W. Norton.

Duarte C, Maurício J, Pettitt PB, Souto P, Trinkaus E, van der Plicht H and Zilháo J (1999) The early Upper Paleolithic human skeleton from the Abrigo do Lagar Velho (Portugal) and modern human emergence in Iberia. *Proceedings of the National Academy of Sciences, USA* 96:7604–7609.

Feder KL (2000) *The Past in Perspective: An Introduction to Human Prehistory.* Second edition. Mountain View, CA: Mayfield.

Feder KL (2002) *Frauds, Myths, and Mysteries: Science and Pseudoscience in Archaeology.* Fourth edition. New York: McGraw-Hill.

Feder KL and Park MA (1997) *Human Antiquity: An Introduction to Physical Anthropology and Archaeology.* Third edition. Mountain View, CA: Mayfield.

Fix AG (1999) *Migration and Colonization in Human Microevolution.* Cambridge: Cambridge University Press.

Foster EA, Jobling MA, Taylor PG, Donnelly P, de Knijff P, Mieremet R, Zerjal T and Tyler-Smith C (1998) Jefferson fathered slave's last child. *Nature* 396:27–28.

Foster EA, Jobling MA, Taylor PG, Donnelly P, de Knijff P, Mieremet R, Zerjal T and Tyler-Smith C (1999) The Thomas Jefferson paternity case. *Nature* 397:32.

Gabunia L, Vekua A , Lordkipanidze D, Swisher III CC, Ferring R, Justus A, Nioradze M, Tvalchrelidze M, Antón G, Bosinski G, Jöris O, de Lumley MA, Majsuradze G and Mouskhelishvili A (2000) Earliest Pleistocene hominid cranial remains from Dmanisi, Republic of Georgia: Taxonomy, geological setting, and age. *Science* 288:1019–1025.

Gagneux P, Wills C, Gerloff U, Tautz D, Morin PA, Boesch C, Fruth B, Hohmann G, Ryder OA and Woodruff DS (1999) Mitochondrial sequences show diverse evolutionary histories of African hominoids. *Proceedings of the National Academy of Sciences, USA* 96:5077–5082.

Gibbons A (2001) The peopling of the Pacific. *Science* 291:1735–1737.

Gibbs S, Collard M and Wood B (2000) Soft-tissue characters in higher primate phylogenetics. *Proceedings of the National Academy of Sciences, USA* 97:11130–11132.

Gmelch G (1977) *The Irish Tinkers: The Urbanization of an Itinerant People.* Menlo Park, CA: Cummings.

Goodman M and Cronin JE (1982) Molecular anthropology: Its development and current directions. In *A History of American Physical Anthropology, 1930–1980,* ed. by F Spencer. New York: Academic Press, pp. 105–146.

Gould SJ (1999) *Rock of Ages: Science and Religion in the Fullness of Life.* New York: Ballantine.

Greenberg JH, Turner CG II and Zegura SL (1986) The settlement of the Americas: A comparison of the linguistic, dental, and genetic evidence. *Current Anthropology* 27:477–497.

Hackett E and Folan ME (1958) The ABO and Rh blood groups of the Aran Islands. *Irish Journal of Medical Science*, June 1958, pp. 247–261.

Haddon AC and Browne CR (1893) The ethnography of the Aran Islands, County Galway. *Proceedings of the Royal Irish Academy* 2:768–830.

Hagelberg E and Clegg JC (1993) Genetic polymorphisms in prehistoric Pacific islanders determined by analysis of ancient bone. *Proceedings of the Royal Society of London B* 252:163–170.

Haile-Selassie Y (2001) Late Miocene hominids from the Middle Awash, Ethiopia. *Nature* 412:178–181.

Hammer MF, Karafet T, Rasanayagam A, Wood ET, Altheide TK, Jenkins T, Griffiths RC, Templeton AR and Zegura SL (1998) Out of Africa and back again: Nested cladistic analysis of human Y chromosome variation. *Molecular Biology and Evolution* 15:427–441.

Hammer MF, Redd AJ, Wood ET, Bonner MR, Jarjanazi H, Karafet T, Santachiara-Benerecetti S, Oppenheim A, Jobling MA, Jenkins T, Orstrer H and Bonné-Tamir B (2000) Jewish and Middle Eastern non-Jewish populations share a common pool of Y-chromosome biallelic haplotypes. *Proceedings of the National Academy of Sciences, USA* 97:6769–6774.

Harding RM, Fullerton SM, Griffiths RC, Bond J, Cox MJ, Schneider JA, Moulin DS and Clegg JB (1997) Archaic African *and* Asian lineages in the genetic ancestry of modern humans. *American Journal of Human Genetics* 60:772–789.

Harpending HC, Batzer MA, Gurven M, Jodre LB, Rogers AR and Sherry ST (1998) Genetic traces of ancient demography. *Proceedings of the National Academy of Sciences, USA* 95:1961–1967.

Harpending HC and Jenkins T (1973) Genetic distance among southern African populations. In *Methods and Theories of Anthropological Genetics*, ed. by MH Crawford and PL Workman. Albuquerque: University of New Mexico Press, pp. 177–199.

Harpending HC, Sherry ST, Rogers AR and Stoneking M (1993) The genetic structure of ancient human populations. *Current Anthropology* 34:483–496.

Harris EE and Hey J (1999) X chromosome evidence for ancient human histories. *Proceedings of the National Academy of Sciences, USA* 96:3320–3324.

Hassan FA (1981) *Demographic Archaeology*. New York: Academic Press.

Hawks J, Oh S, Hunley K, Dobson S, Cabana G, Dayalu P and Wolpoff MH (2000) An Australasian test of the recent African origin model using the WLH-50 calvarium. *Journal of Human Evolution* 39:1–22.

Hill EW, Jobling MA and Bradley DG (2000) Y-chromosome variation and Irish origins. *Nature* 404:351.

Hooton EA, Dupertuis CW and Dawson H (1955) *The Physical Anthropology of Ireland*. Papers of the Peabody Museum, Vol. 30, Nos. 1–2. Cambridge, MA: Peabody Museum.

Horai S, Hayasaka K, Kondo R, Tsugane K and Takahata N (1995) Recent African origin of modern humans revealed by complete sequences of hominoid mitochondrial DNAs. *Proceedings of the National Academy of Sciences, USA* 92:532–536.

Horai S, Kondo R, Nakagawa-Hattori Y, Hayashi S, Sonoda S and Tajima K (1993) Peopling of the Americas, founded by four major lineages of mitochondrial DNA. *Molecular Biology and Evolution* 10:23–47.

Howell FC (1952) Pleistocene glacial ecology and the evolution of "Classic Neandertal" man. *Southwestern Journal of Anthropology* 8:377–410.

Howells WW (1989) *Skull Shapes and the Map: Craniometric Analyses in the Dispersion of Modern Homo.* Papers of the Peabody Museum of Archaeology and Ethnology, Volume 79. Cambridge, MA: Harvard University Press.

Kaessmann H, Wiebe V and Pääbo S (1999) Extensive nuclear DNA sequence diversity among chimpanzees. *Science* 286:1159–1162.

Karafet TM, Zegura SL, Posukh O, Osipovsa L, Bergen A, Long J, Goldman D, Klitz W, Harihara S, de Knijff P, Wiebe V, Griffiths RC, Templeton AR and Hammer MF (1999) Ancestral Asian source(s) of New World Y-chromosome founder haplotypes. *American Journal of Human Genetics* 64:817–831.

Kayser M, Brauer S, Weiss G, Underhill PA, Roewer L, Schiefenhövel W and Stoneking M (2000) Melanesian origin of Polynesian Y chromosomes. *Current Biology* 10:1237–1246.

Kennedy KAR (1976) *Human Variation in Time and Space.* Dubuque, IA: Wm. C. Brown.

King MC and Wilson AC (1975) Evolution at two levels in humans and chimpanzees. *Science* 188:107–116.

King W (1864) The reputed fossil man of the Neanderthal. *Quarterly Journal of Science* 1:88–97.

Kobyliansky E, Micile S, Goldschmidt-Nathan M, Arensburg B and Nathan H (1982) Jewish populations of the world: Genetic likeness and differences. *Annals of Human Biology* 9:1–34.

Krings M, Capelli C, Tschentscher F, Geisert H, Meyer S, von Haeseler A, Grossschmidt K, Possnert G, Paunovic M and Pääbo S (2000) A view of Neandertal genetic diversity. *Nature Genetics* 26:144–146.

Krings M, Geisert H, Schmitz RW, Krainitzki H and Pääbo S (1999) DNA sequence of the mitochondrial hypervariable region II from the Neandertal type specimen. *Proceedings of the National Academy of Sciences, USA* 96:5581–5585.

Krings M, Stone A, Schmitz RW, Krainitzki H, Stoneking M and Pääbo S (1997) Neandertal DNA sequences and the origin of modern humans. *Cell* 90:19–30.

Lahr MM (1996) *The Evolution of Modern Human Diversity: A Study of Cranial Variation.* Cambridge: Cambridge University Press.

Lander ES and Ellis JJ (1998) Founding father. *Nature* 396:13–14.

Lewin R (1997) *Bones of Contention: Controversies in the Search for Human Origins.* Second edition. Chicago: University of Chicago Press.

Livshits G, Sokal RR and Kobyliansky E (1991) Genetic affinities of Jewish populations. *American Journal of Human Genetics* 49:131–146.

Lum JK, Cann RL, Martinson JJ and Jorde LB (1998) Mitochondrial and nuclear genetic relationships among Pacific Island and Asian populations. *American Journal of Human Genetics* 63:613–624.

Malhi RS, Eshleman JA, Greenberg JA, Weiss DA, Schultz Shook BA, Kaestle FE, Lorenz JG, Kemp BM, Johnson JR and Smith DG (2002) The structure of diversity within New World mitochondrial DNA haplogroups: Implications for the prehistory of North America. *American Journal of Human Genetics* 70:905–919.

Marks J (1994) Black, white, other. *Natural History* 103:32–35.

Marshall E (2001) Pre-Clovis sites fight for acceptance. *Science* 291:1730–1732.

Matisoo-Smith E, Roberts RM, Allen JS, Penny D and Lambert DM (1998) Patterns of prehistoric human mobility in Polynesia indicated by mtDNA from the Pacific rat. *Proceedings of the National Academy of Sciences, USA* 95:15145–15150.

McKean K (1983) Preaching the molecular gospel. *Discover* 4(7):34–40.

Meltzer DJ (1997) Monte Verde and the Pleistocene peopling of the Americas. *Science* 276:754–755.

Menozzi P, Piazza A and Cavalli-Sforza LL (1978) Synthetic maps of human gene frequencies in Europe. *Science* 201:786–792.

Merriwether DA, Friedlaender JS, Mediavilla J, Mgone C, Gentz F and Ferrell RE (1999) Mitochondrial DNA variation is an indicator of Austronesian influence in Island Melanesia. *American Journal of Physical Anthropology* 110:243–270.

Merriwether DA, Huston S, Iyengar S, Hamman R, Norris JM, Shetterly SM, Kamboh MI and Ferrell RE (1997) Mitochondrial versus nuclear admixture estimates demonstrate a past history of directional mating. *American Journal of Physical Anthropology* 102:153–159.

Merriwether DA, Rothhammer F and Ferrell RE (1995) Distribution of the four founding lineage haplotypes in Native Americans suggests a single wave of migration for the New World. *American Journal of Physical Anthropology* 98:411–430.

Morris D (1967) *The Naked Ape*. New York: Dell.

Mourant AE, Kopec AC and Domaniewska-Sobczak K (1976) *The Distribution of the Human Blood Group and Other Polymorphisms*. Oxford: Oxford University Press.

Nemecek S (2000) Who were the first Americans? *Scientific American* 283(3):80–88.

North KE, Martin LJ and Crawford MH (2000) The origins of the Irish travelers and the genetic structure of Ireland. *Annals of Human Biology* 27:453–465.

Ovchinnikov I, Götherstrom A, Romanova GP, Kharitonov VM, Lidén K and Goodwin W (2000) Molecular analysis of Neanderthal DNA from the northern Caucasus. *Nature* 404:490–493.

Parra EJ, Kittles RA, Argyropoulos G, Pfaff CL, Hiester K, Bonilla C, Sylvester N, Parrish-Gause D, Garvey WT, Jin L, McKeigue PM, Kamboh MI, Ferrell RE, Pollitzer WS and Shriver MD (2001) Ancestral proportions and admixture dynamics in geographically defined African Americans living in South Carolina. *American Journal of Physical Anthropology* 114:18–29.

Parra EJ, Marcini A, Akey J, Martinson J, Batzer MA, Cooper R, Forrester T, Allison DB, Deka R, Ferrell RE and Shriver MD (1998) Estimating African American admixture proportions by use of population-specific alleles. *American Journal of Human Genetics* 63:1839–1851.

Penny D, Steel M, Waddell PJ and Hendy MD (1995) Improved analyses of human mtDNA sequences support a recent African origin for *Homo sapiens*. *Molecular Biology and Evolution* 12:863–882.

Piazza A, Rendine S, Mich E, Menozzi P, Mountain J and Cavalli-Sforza LL (1995) Genetics and the origin of European languages. *Proceedings of the National Academy of Sciences, USA* 92:5836–5840.

Population Reference Bureau (2001) *2001 World Population Data Sheet*. Washington, DC. Available on the Internet at: http://www.prb.org/pubs/wpds2001/

Powell JF and Neves WA (1999) Craniofacial morphology of the first Americans: Pattern and process in the peopling of the New World. *Yearbook of Physical Anthropology* 42:153–188.

Price TD and Feinman GM (2001) *Images of the Past.* Third edition. Mountain View, CA: Mayfield.

Redd AJ, Takezaki N, Sherry ST, McGarvey ST, Sofro ASM and Stoneking M (1995) Evolutionary history of the COII/tRNA^{Lys} intergenic 9 base pair deletion in human mitochondrial DNAs from the Pacific. *Molecular Biology and Evolution* 12:604–615.

Reed TE (1969) Caucasian genes in American Negroes. *Science* 165:762–768.

Relethford JH (1982) Isonymy and population structure of Irish isolates during the 1890s. *Journal of Biosocial Science* 14:241–247.

Relethford JH (1988) Effects of English admixture and geographic distance on anthropometric variation and genetic structure in 19th-century Ireland. *American Journal of Physical Anthropology* 76:111–124.

Relethford JH (1991) Genetic drift and anthropometric variation in Ireland. *Human Biology* 63:155–165.

Relethford JH (1994) Craniometric variation among modern human populations. *American Journal of Physical Anthropology* 95:53–62.

Relethford JH (1997) Hemispheric difference in human skin color. *American Journal of Physical Anthropology* 104:449–457.

Relethford JH (2001a) Absence of regional affinities of Neandertal DNA with living humans does not reject multiregional evolution. *American Journal of Physical Anthropology* 115:95–98.

Relethford JH (2001b) Ancient DNA and the origin of modern humans. *Proceedings of the National Academy of Sciences, USA* 98:390–391.

Relethford JH (2001c) *Genetics and the Search for Modern Human Origins.* New York: John Wiley & Sons.

Relethford JH (2002) Apportionment of global human genetic diversity based on craniometrics and skin color. *American Journal of Physical Anthropology* 118:393–398.

Relethford JH (2003) Anthropometric data and population history. In *Human Biologists in the Archives,* ed. by A Herring and AC Swedlund. Cambridge: Cambridge University Press, pp. 32–52.

Relethford JH and Blangero J (1990) Detection of differential gene flow from patterns of quantitative variation. *Human Biology* 62:5–25.

Relethford JH and Crawford MH (1995) Anthropometric variation and the population history of Ireland. *American Journal of Physical Anthropology* 96:25–38.

Relethford JH and Harpending HC (1994) Craniometric variation, genetic theory, and modern human origins. *American Journal of Physical Anthropology* 95:249–270.

Relethford JH and Jorde LB (1999) Genetic evidence for larger African population size during recent human evolution. *American Journal of Physical Anthropology* 108:251–260.

Richards M, Córte-Real H, Forster P, Macaulay V, Wilkinson-Herbots H, Demaine A, Papiha S, Hedges R, Bandelt H-J and Sykes B (1996) Paleolithic and Neolithic lineages in the European mitochondrial gene pool. *American Journal of Humans Genetics* 59:185–203.

Richards M, Oppenheimer S and Sykes B (1998) mtDNA suggests Polynesian origins in Eastern Indonesia. *American Journal of Human Genetics* 63:1234–1237.

Ritte U, Neufeld E, Prager EM, Gross M, Hakim I, Khatib A and Bonné-Tamir B (1993) Mitochondrial DNA affinity of several Jewish communities. *Human Biology* 65:359–385.

Roychoudhury AK and Nei M (1988) *Human Polymorphic Genes: World Distribution*. New York: Oxford University Press.

Ruvolo M, Zehr S, von Dornum M, Pan D, Chang B and Lin J (1993) Mitochondrial COII sequences and modern human origins. *Molecular Biology and Evolution* 10:1115–1135.

Sarich VM (1971) A molecular approach to the question of human origins. In *Background for Man: Readings in Physical Anthropology*, ed. by P Dohlinow and VM Sarich. Boston: Little, Brown, pp. 60–81.

Sarich V and Wilson A (1967) Immunological time scale for hominoid evolution. *Science* 158:1200–1203.

Schafer D (1996a) Ancient Tri-Citian's bones raise controversy. *Tri-City Herald*, September 9, 1996. Available at http://www.kennewick-man.com/news/0909.html

Schafer D (1996b) Skull likely early white settler. *Tri-City Herald*, July 30, 1996. Available at http://www.kennewick-man.com/news/073096.html

Schurr TG (2000) Mitochondrial DNA and the peopling of the New World. *American Scientist* 88:246–253.

Senut B, Pickford M, Gommery D, Mein P, Cheboi K and Coppens Y (2001) First hominid from the Miocene (Lukeino formation, Kenya). *Comptes Rendus de l'Académie des Sciences, Paris* 332:137–144.

Smith FH, Trinkaus E, Pettitt PB, Karavaniæ I and Paunoviæ M (1999) Direct radiocarbon dates for Vindija G1 and Velika Peæina Late Pleistocene hominid remains. *Proceedings of the National Academy of Sciences, USA* 96:12281–12286.

Sokal RR, Oden NL, Rosenberg MS and Di Giovanni D (1997) The patterns of historical population movements in Europe and some of their genetic consequences. *American Journal of Human Biology* 9:391–404.

Sokal RR, Oden NL and Thompson BA (1992) Origins of the Indo-Europeans: Genetic evidence. *Proceedings of the National Academy of Sciences, USA* 89:7669–7673.

Sokal RR, Oden NL, Walker J, Di Giovanni D and Thompson BA (1996) Historical population movements in Europe influence genetic relationships in modern samples. *Human Biology* 68:873–898.

Sokal RR and Rohlf FJ (1995) *Biometry: The Principles and Practice of Statistics in Biological Research*. Third edition. New York: W. H. Freeman.

Spurdle AB and Jenkins T (1996) The origins of the Lemba "Black Jews" of southern Africa: Evidence from p12F2 and other Y-chromosome markers. *American Journal of Human Genetics* 59:1126–1133.

Stang J (1996a) Many claim to have connection to Kennewick Man. *Tri-City Herald*, December 20, 1996. Available at http://www.kennewick-man.com/news/12201.html

Stang J (1996b) Tri-City skeleton dated at 9,000 years old. *Tri-City Herald*, August 28, 1996. Available at http://www.kennewick-man.com/news/0828.html

Stoneking M (1993) DNA and recent human evolution. *Evolutionary Anthropology* 2:60–73.

Stringer C and Gamble C (1993) *In Search of the Neanderthals*. New York: Thames and Hudson.

Stringer C and McKie R (1996) *African Exodus: The Origins of Modern Humanity.* New York: Henry Holt.

Su B, Jin L, Underhill P, Martinson J, Saha N, McGarvey ST, Shriver MD, Chu J, Oefner P, Chakraborty R and Deka R (2000) Polynesian origins: Insights from the Y chromosome. *Proceedings of the National Academy of Sciences, USA* 97:8225–8228.

Swisher CC, Curtis GH, Jacob T, Getty AG, Suprijo A and Widiasmoro (1994) Age of the earliest known hominids in Java, Indonesia. *Science* 263:1118–1121.

Sykes B (2001) *The Seven Daughters of Eve*. New York: W. W. Norton.

Sykes B, Leiboff A, Low-Beer J, Tetzner S and Richards M (1995) The origins of the Polynesians: An interpretation from mitochondrial lineage analysis. *American Journal of Human Genetics* 57:1463–1475.

Szathmary EJE (1993) Genetics of aboriginal North Americans. *Evolutionary Anthropology* 1:202–220.

Tattersall I (1994) How does evolution work? *Evolutionary Anthropology* 3:2–3.

Tattersall I (1995) *The Fossil Trail: How We Know What We Think We Know About Human Evolution*. New York: Oxford University Press.

Tattersall I and Schwartz JH (1999) Hominids and hybrids: The place of Neanderthals in human evolution. *Proceedings of the National Academy of Sciences, USA* 96:7117–7119.

Templeton AR (1993) The "Eve" hypothesis: A genetic critique and reanalysis. *American Anthropologist* 95:51–72.

Templeton AR (1997) Testing the Out of Africa replacement hypothesis with mitochondrial DNA data. In *Conceptual Issues in Modern Human Origins Research,* ed. by GA Clark and CM Willermet. New York: Aldine de Gruyter, pp. 329–360.

Templeton AR (1998) Human races: A genetic and evolutionary perspective. *American Anthropologist* 100:632–650.

Templeton AR (2002) Out of Africa again and again. *Nature* 416:45–51.

Terrell J (1988) History as a family tree, history as an entangled bank: Constructing images and interpretations of prehistory in the South Pacific. *Antiquity* 62:642–657.

Terrell J (2001) Foregone conclusions? In search of "Papuans" and "Austronesians." *Current Anthropology* 42:97–124.

Thomas DH (2000) *Skull Wars: Kennewick Man, Archaeology, and the Battle for Native American Identity.* New York: Basic Books.

Thomas MG, Parfitt T, Weiss DA, Skorecki K, Wilson JF, le Roux M, Bradman N and Goldstein DB (2000) Y chromosomes traveling south: The Cohen modal haplotypes and the origins of the Lemba—the "Black Jews of Southern Africa." *American Journal of Human Genetics* 66:674–686.

Thomas MG, Skorecki K, Ben-Ami H, Parfitt T, Bradman N and Goldstein DB (1998) Origins of Old Testament priests. *Nature* 394:138–140.

Thomas MG, Weale ME, Jones AL, Richard M, Smith A, Redhead N, Torroni A, Scozzari R, Gratrix F, Tarekegn A, Wilson JF, Capelli C, Bradman N and Goldstein DB (2002) Founding mothers of Jewish communities: Geographically separated Jewish groups were independently founded by very few female ancestors. *American Journal of Human Genetics* 70:1411–1420.

Thorne AG, Wolpoff MH and Eckhardt RB (1993) Genetic variation in Africa. *Science* 261:1507–1508.

Tills D, Teesdale P and Mourant AE (1977) Blood groups of the Irish. *Annals of Human Biology* 4:23–34.

Torroni A, Bandelt H-J, D'Urbano L, Lahermo P, Moral P, Sellito D, Rengo C, Forster P, Savontaus M-L, Bonné-Tamir B and Scozzari R (1998) mtDNA analysis reveals a major Late Paleolithic population expansion from southwestern to northeastern Europe. *American Journal of Human Genetics* 62:1137–1152.

Trinkaus E and Shipman P (1992) *The Neandertals: Changing the Image of Mankind.* New York: Alfred A. Knopf.

Underhill PA, Passarino G, Lin AA, Marzuki S, Oefner PJ, Cavalli-Sforza LL and Chambers GK (2001) Maori origins, Y-chromosome haplotypes and implications for human history in the Pacific. *Human Mutation* 17:271–280.

Vaughan WE and Fitzpatrick AJ (1978) *Irish Historical Statistics: Population, 1821–1971.* Dublin: Royal Irish Academy.

Wainscoat J (1987) Out of the Garden of Eden. *Nature* 325:13.

Weiss ML and Mann AE (1990) *Human Biology and Behavior: An Anthropological Perspective.* Fifth edition. Glenview, IL: Scott, Foresman/Little, Brown.

Whitlock MC and Barton NH (1997) The effective size of a subdivided population. *Genetics* 146:427–441.

Wolpoff MH (1994) How does evolution work? *Evolutionary Anthropology* 3:4–5.

Wolpoff MH (1999) *Paleoanthropology.* Second edition. Boston: McGraw-Hill.

Wolpoff MH and Caspari R (1997) What does it mean to be modern? In *Conceptual Issues in Modern Human Origins Research,* ed. by GA Clark and CM Willermet. New York: Aldine de Gruyter, pp. 28–44.

Wolpoff MH, Hawks J, Frayer DW and Hunley K (2001) Modern human ancestry at the peripheries: A test of the replacement theory. *Science* 291:293–297.

Woodham-Smith C (1962) *The Great Hunger: Ireland, 1845–1849.* New York: Harper & Row.

Wright S (1969) *Evolution and the Genetics of Populations, Volume 2: The Theory of Gene Frequencies.* Chicago: University of Chicago Press.

INDEX